THE
HANDICAP
PRINCIPLE

THE HANDICAP PRINCIPLE

A Missing Piece of Darwin's Puzzle

AMOTZ AND AVISHAG ZAHAVI

with Naama Zahavi-Ely and Melvin Patrick Ely
Illustrations by Amir Balaban

OXFORD UNIVERSITY PRESS
NEW YORK   OXFORD

Oxford University Press

Oxford   New York

Athens   Auckland   Bangkok   Bogotá   Buenos Aires   Calcutta
Cape Town   Chennai   Dar es Salaam   Delhi   Florence   Hong Kong
Istanbul   Karachi   Kuala Lumpur   Madrid   Melbourne   Mexico City
Mumbai   Nairobi   Paris   São Paolo   Singapore   Taipei   Tokyo   Toronto   Warsaw
and associated companies in
Berlin   Ibadan

Library of Congress Cataloging-in-Publication Data
Zahavi, Amotz.
The handicap principle : a missing piece of Darwin's puzzle / by
Amotz and Avishag Zahavi ; translation by Naama Zahavi-Ely and
Melvin Patrick Ely.
p. cm.
Translated from Hebrew.
Includes bibliographical references and index.
ISBN 0-19-510035-2
ISBN 0-19-512914-8 (Pbk.)
1. Animal behavior.   2. Animal communication   I. Zahavi,
Avishag.   II. Title.
QL751.Z44   1997   96-42374
591.59—DC20

1 3 5 7 9 10 8 6 4 2

Printed in the United States of America
on acid-free paper

Illustrations by Amir Balaban

CONTENTS

Alarm calls: a message to friends or to foes? Stotting as communication with predators. Calls by prey during pursuit: the merlin and the skylark. Warning colors (aposematic coloration). Signaling to prey. Cooperation between prey and predators without communication. Preconditions for prey-predator communication.

Threats as a substitute for aggression: are humans worse than beasts? Handicaps as keys for reliable threats. Threats by approaching. Threats by stretching—deceit or exposure to danger? Threats by vocalization. Other threats. Social hierarchies and duels between equals. Can threat signals evolve for the good of the group? The drawbacks of "group selection."

The conflict inherent in courtship. Courtship handicaps and the information they convey: feeding ability, superterritories, courtship vocalization, colors, scents (pheromones), artifacts and constructions, combinations of signals, long tails, movements and dances. Leks: congregating for display. Polymorphic species and males that mimic females. Fisher's model of the Runaway Process compared with the Handicap Principle. Utilitarian selection and signal selection.

PART II METHODS OF COMMUNICATION

Did decoration evolve to identify species, gender, and age? The evolution of common markings through competition between individuals. Markings and the traits they advertise: lines and stripes, patches and frames that bring out body parts. Implications of the Handicap Principle: the use of markings to focus on features. Symmetry. "Eye" patterns. The evolution of markings; polymorphism and convergence. Facial markings and the direction of gaze. Status badges or handicaps? Are there signals without handicaps? Are there any signals that are conventions? The definition of signals; inflation as a test of the theory of signal selection.

Difficult movements. Ritual fighting. Ritualization: does it reduce the information conveyed? How a ritualized signal evolves.

The correlation of voice with posture and tension. The information conveyed by vocal signals. Animal vocabularies: the connection between the message and its vocal pattern. Rhythm. Vocal patterns used over distance. Why shout? The duration of vocalization: requests and commands. Dialogues and their significance. Mimicry. Do animals have verbal language?

Long tails: are they signals? Bristling hair or feathers: an illusion of size, or a handicap? Manes and crests. Handicaps that interfere with vision. Body parts that emphasize the direction of gaze. Body parts that handicap fighting. Can body parts evolve to reduce the cost of signals? The evolution of horns and antlers. Signal selection and the evolution of feathers.

Black in the desert. Black and white in open spaces. Colors in forests and on coral reefs. The use of two colors. Glossy colors and movement. Exceptions to the rules.

Pheromones in butterflies and moths: chemical handicaps. Yeast sex pheromones and propheromones: the role of glycoproteins. Chemical communication within the multicellular body.

Testing by imposition. Aggression in courtship. Hide-and-seek: gentler testing in courtship. Clumping, and preening others (allopreening). Group dances and similar rituals.

Threats of self-injury: the weapon of the weaker partner. Other methods of blackmail. Exploitation of offspring by parents.

Territories, groups, and nonterritorial individuals. Rank, avoidance of incest, and the life strategy of males and of females. The composition of groups; coalitions of males and of females. Struggles without aggression (well, almost). Theories that explain altruism: group-selection theory and its failings. The theory of reciprocal altruism and the problem of enforcement. Competition over altruistic acts in babblers: sentinel activities, feeding of nestlings, feeding of other adults (allofeeding), mobbing. Altruism as a substitute for threats. Rank and prestige. "Shyness" over copulation as a test of male prestige. Reasons for and consequences of living in groups. Prestige and the evolution of altruism: altruism as a handicap.

The evolution of social structures in the social insects: conditions that favor collaboration: food storage and helpless offspring; the haplodiploid mechanism of gender determination; selection through queens only, or selection through workers too? How insect colonies form. Why do the workers work for the colony? altruism and prestige; queen pheromones and prestige; the handicap in queen pheromone. Kin selection theory and its drawbacks: parasitism among kin, or Haldane's other brothers; are offspring "kin"? Partnerships among kin: why it makes sense to join the family business. The kin effect.

Paternity and mate-guarding. Taking care of the young to gain prestige. Other means of showing off to one's mate. Dominance between mates. Conditions for female dominance. The parental couple as a partnership.

The life cycle of cellular slime molds. Forming the stalk: altruistic suicide? The individual selection hypothesis. DIF as a poison. The difference between prestalk and prespore amebas. Some remaining questions. When is a chemical a signal?

An arms race or a state of equilibrium? European cuckoos and reed warblers. Great spotted cuckoos and crows. The prestige model. The Mafia model. Accepting a parasite to minimize damage. Neutering the host. From parasite to collaborator. The less virulent parasite as a collaborator against its virulent variant. The implications of assuming a state of equilibrium.

Food sources and social organization: the white wagtail. Communal roosts as information centers. Insurance against evil days: winter gatherings of rooks. Flocks and loners: the communal roost of kites in Coto Doñana. Bright adults and dull youngsters: handicaps in food squabbles. How information repositories work. Human gatherings. Communal displays at gatherings: promoting the roost or mutual testing?

Innate behavior in humans. The human body and its decorations: hair; eyes, eyebrows, and eyelashes; nose and facial wrinkles; chin and beard;

red cheeks and lips; menstruation; breasts and body fat; clothing. Testing the human social bond: the human sexual act as a test of the bond, self-endangerment in humans: suicide as a cry for help. Human language: communication without reliability. Decoration, esthetics, and the evolution of art. Altruism and moral behavior.

ACKNOWLEDGMENTS

Our daughters Naama and Tirza took part from the very beginning in the discussions from which our ideas emerged, and their suggestions have helped us present those ideas here. Naama's presence in these discussions, and that of her husband Melvin Patrick Ely in later years, equipped the two of them, working from our original Hebrew text, to write the version of this book that now appears in English and other languages. We also thank Amir Balaban for the skill and artistry he brought to the task of illustrating the book.

Azaria Alon encouraged and helped us throughout, but especially with the difficult early steps. We wish to thank Daniela Atzmony, Helena Cronin, Paul Eckman, Michal Gil, Jehoshua Kugler, Arnon Lotem, Jonathan Wright, Mina Yarom, Yoram Yom-Tov, and Zohar Zuk-Rimon for reading and commenting on the whole manuscript or on some of its chapters.

The first drafts of this book were written well over ten years ago. It is impossible for us to acknowledge all our friends, students, and colleagues who read some of these drafts and made important comments and suggestions for improvements, or those who provided photographs or videocassettes, on which Amir Balaban has based some of his lively drawings. We do hope that these friends will forgive us for not mentioning all of them by name.

We thank the staff of Oxford University Press, especially our editor Kirk Jensen, our production editor Kimberly Torre-Tasso, and our devoted copyeditor, Nora Cavin. We gratefully acknowledge the efforts of our literary agent, Richard Balkin.

Special thanks are due to the many volunteers and students who helped with our field observations of the babblers, getting to know these cooperatively living birds individually and recording data about their behavior. But above all we thank the Society for the Protection of Nature in Israel (SPNI) for allowing all of us to live and work in its Field Study Center at Hatzeva.

The income from the Hebrew edition of this book is dedicated to a fund which will be used to continue our study of the babblers and to maintain the Shezaf Nature Reserve, where these birds can continue to live, protected from the upheavals created by modern life. We hope that the English version and other translations of this book will widen the circle of friends of the Shezaf Nature Reserve and its babblers.

Amotz and Avishag Zahavi

THE GAZELLE, THE WOLF, AND THE PEACOCK'S TAIL

We start with a scene of a gazelle resting or grazing in the desert. It is nearly invisible; the color of its coat blends well with the desert landscape. A wolf appears. One would expect the gazelle to freeze or crouch and do its utmost to avoid being seen. But no: it rises, barks, and thumps the ground with its forefeet, all the while watching the wolf. The thumps of the gazelle's hooves carry through the desert ground over long distances; its curved horns and the dark-and-light pattern on its face clearly reveal that the gazelle is in fact looking at its enemy.

If the wolf comes nearer, one would expect the gazelle to flee as fast as it can. But no again: often the gazelle jumps high on all four legs several times and only then begins to run, wagging its short black tail against its conspicuous white rump, which has a black border. These high jumps are very clearly linked to the approach of the wolf. Yet a gazelle escaping immediate, urgent danger—such as hunters in a jeep—flees in an entirely different manner: it runs away silently at great speed, making good use of the topography to conceal its escape.

Why does the gazelle reveal itself to a predator that might not otherwise spot it? Why does it waste time and energy jumping up and down (*stotting*) instead of running away as fast as it can? The gazelle is signaling to the predator that it has seen it; by "wasting" time and by jumping high in the air rather than bounding away, it demonstrates in a reliable way that it is able to outrun the wolf. The wolf, upon learning that it has lost its chance to surprise its prey, and that this gazelle is in tip-top physical shape, may decide to move on to another area; or it may decide to look for more promising prey.

Even parties in the most adversarial relationships, such as prey and predator, may communicate, provided that they have a common interest: in this case, both want to avoid a pointless chase. The gazelle tries to convince the wolf that it is not the easy prey the wolf is looking for, and that the wolf would be wasting time and energy by chasing it. Even if the gazelle is sure that it can outrun the wolf, it too would prefer to avoid an exhausting chase. But in order to convince the wolf not to give chase, the gazelle has to expend precious time and energy that it will need should the wolf ignore its signals and decide to chase it anyway.

The encounter between the gazelle and the wolf dramatizes the basic theme of this book: in order to be effective, signals have to be reliable; in order to be reliable, signals have to be costly.

The high cost that animal signaling often involves is clearly seen in the case of the peacock. Most people have seen and admired a peacock, spreading and quivering his enormous tail—a fan of glistening feathers, adorned with blue and green "eyes." But to be able to put on such shows, peacocks have to drag massive tails around most of the year. By managing to find food and avoid predators despite such a burden, a peacock proves that he is the high-quality mate that the peahen is seeking to father her future chicks.

This is another basic theme of our book: that there is a logical relationship between the signal and the message it conveys. The gazelle displays its confidence in its ability to outrun the predator by drawing attention to itself and by expending precious time and energy that it will need should its signal not be heeded. The peacock proves his strength and agility by carrying a heavy load, as does a stag carrying heavy antlers. Each signal is closely related to its message. A person can signal courage by courting danger, for example, but courting danger does not attest to wealth—which may be demonstrated by wasting money.

The investment that animals make in signals is similar to the "handicaps" imposed on the stronger contestants in a game or a sporting event: for example, the removal of the superior player's queen in a chess match, the extra weight the swifter race horse must carry, or the score of several strokes that the more accomplished golfer starts with. A handicap proves beyond a doubt that the victor's win is due to mastery, not chance. The peacock's tail and the stag's antlers are not mere disabilities; rather, they are handicaps in this very special sense: they allow an individual animal to demonstrate its quality.

Are animal signals always reliable? We believe that most signals are: before an individual acts on information it receives through signals from another individual, it first needs to check the reliability of that information. We suggest a very simple principle: if a given signal requires the signaler to invest more in the signal than it would gain by conveying phony information, then faking is unprofitable and the signal is therefore credible. If a gazelle that is slow or weak sends the wolf a phony signal about its speed and strength by stotting, it wastes what little strength it has on puny jumps that will only convince the wolf that it is easy prey. Such a gazelle would do better to flee for its life and hope for the best. To gauge the reliability of a signal, then, one has to consider the investment it entails. The cost—the handicap that the signaler takes on—guarantees that the signal is reliable.

When we first suggested this Handicap Principle in 1975[1] it was almost unanimously rejected. Many papers were published using formal, explicit mathematical models "proving" that the Handicap Principle does not work,[2] or that it might apply only under very special conditions.[3] This trend changed in 1990 when Alan Grafen published two papers[4] using different mathematical models to show that the Handicap Principle is generally applicable, and that it is a sound principle that can ensure reliability in communication between competing organisms. Since then, the Handicap Principle has become widely accepted.[5]

Throughout all these years, while our colleagues were debating the validity of the principle, we continued to observe and explore the living world around us. The Handicap Principle revealed to us an endless array of new ways to understand such phenomena as the extreme expenditures often involved in sexual advertisement, the evolutionary enigma of animal altruism, and the workings of collaborative systems in the animal kingdom, which could not easily be explained in terms of straightforward, utilitarian natural selection.

Our investigations into the ramifications of the Handicap Principle coincided with our studies of the babblers, group-living songbirds, at Hatzeva in the Arava Valley of southern Israel. We have been studying these small desert birds for the last twenty-five years. They are used to our presence; we often get within a few feet or even inches of them. As far as they are concerned, we are not much different from the herds of goats and camels with which they share the desert. We can watch them very closely, observe the fine details of their behavior, and hear the soft and subtle vocalizations with which they communicate with one another. We can identify each individual bird by the colored leg bands we put on its legs when it was a nestling.

All members of a babbler group participate in the defense of their territory and in the care of nestlings, even when the nestlings are not their own offspring.

The birds perform many other altruistic acts, such as feeding other adult members of their group and standing guard while the rest of the group is feeding. Such activities are difficult for researchers of evolution to explain. We discovered that these altruistic behaviors serve to advertise each babbler. In other words, each babbler's investment in altruistic acts demonstrates the validity of its claim to social status—to prestige. We learned from babblers how important it is to animals to be able to measure the strength of social bonds; that need explained the babblers' group "dances," and their clumping together to rest. The Handicap Principle also explains the birds' shyness about sex and even the details of their very subtle decorative markings.

We did not start out seeking a unifying principle in biological communication. All we attempted to do at first, in 1973, was to explain the evolution of the peacock's tail to a student and colleague, Yoav Sagi, who—for good reason, as it turned out—could not see the logic in Fisher's runaway process, then the current theory. Our broad application of the Handicap Principle developed slowly: one finding led to another, until we realized we were dealing with a general principle.

Many people, including those who now accept the Handicap Principle, have not yet attempted to apply it broadly to biological signaling. We think that the explanations and models we present in this book are more plausible, and fit the known facts better, than those commonly found in textbooks. We expect our ideas to serve as a starting point for more detailed studies and experiments. Some of our explanations will prove valid as they stand; in other cases, reality will turn out to be even more fascinating and complex than we now imagine. We hope that this book will encourage the search for reliability and for its unavoidable cost in all systems of biological signaling. We believe that this search will change our understanding of the natural world in a myriad of ways.

PART 1

PARTNERS IN COMMUNICATION

CHAPTER 1

PREY–PREDATOR INTERACTIONS

ALARM CALLS: A MESSAGE TO FRIENDS OR TO FOES?

abblers move around their territory during the day, looking for food on the ground among the desert trees and bushes. One of the group often perches on a treetop, acting as sentinel. When the sentinel, or any other babbler, sees a bird of prey, or raptor, in the distance, it emits a loud "bark." Upon hearing such barks, babblers often raise their heads and scan the sky. If the raptor is far away and does not pose any immediate danger to the feeding group, they go on searching for food. But when an abrupt, frightened bark indicates imminent danger, the birds, including the sentinel, immediately jump for cover. Often, however, precisely after such a fright, the entire group goes up to the top of the tree in which they have taken shelter and joins the sentinel in barking and calling loudly.

In most studies of animal behavior such vocalizations are classified as warning calls, issued by animals to others of their species. That idea has a certain superficial plausibility. But years of observation raised so many puzzling questions that we finally dared ask ourselves whether the calls were indeed meant as warnings. The barks ring out before the raptor has had any chance to notice the group or pose any danger to it. The calls seem to be an unjustified risk for the barking birds: a feeding group of drab and well-camouflaged babblers has a good chance of being overlooked by a raptor altogether, yet the barks disclose their location over a considerable distance. In fact, when we try to find babblers in their vast territories, which often cover more than a square kilometer, we have a good chance of missing them when they feed quietly among the bushes; but as soon as we see a raptor in the distance we stop the car and listen for barking babblers. The sharp barks the birds emit periodically as long as a raptor is around allow us—and probably the raptor—to locate the group. The calls seem to serve the interests of the predator rather than the group.

One could possibly understand the first bark as a warning to the group, and perhaps the second and the third as efforts to make all members of the group aware of the danger. But what is the point of repeating the calls after the entire group has already taken cover? And whom are they "warning," when the entire group joins the sentinel at the top of the canopy, barking together?

And if the bark is a warning to the rest of the group, why is it so loud? Babblers often vocalize softly while feeding, but the "warning calls" are hundreds of times louder than these sounds and may be heard over half a mile away. Why do babblers raise the volume of their calls to a level that can alert a predator to their presence precisely when a predator is in the area—and before it has had a chance to notice the well-camouflaged group?

There is also a theoretical difficulty regarding the evolution of warning calls. For a trait—such as the tendency to call out when one spots a predator—to spread through a population by natural selection, that trait has to improve the chances that the specific individual who possesses it will survive and pass it on to descendants. But the investment in so-called warning calls is made by the callers, while the benefit goes to the listeners.[1]

Indeed, callers and listeners are not even necessarily of the same species: at Hatzeva, where we observe, the babblers, shrikes, blackstarts, bulbuls, and wheatears also call out loudly at the appearance of raptors. Some of these birds, such as shrikes in winter, are solitary and have no collaborator to warn, not even a mate. The fact that the barks tell anybody who happens to be around, including other bird species—and for that matter, including us—that there is a raptor in the sky does not mean that they evolved for that purpose. So we searched for an alternative explanation.

Loud calls make sense when the listener is distant.[2] Since the raptor is far away from the sentinel, is it possible that the intended listener is actually the raptor? The notion makes even more sense in view of the callers' behavior after a raptor lands: the babblers often approach the predator and "mob" it. One does not hear any clear separation between the warning calls and the mobbing calls: one behavior merges into the other. It is widely accepted that mobbing is directed at the predator[3]; is it possible that the "warning" calls are also directed at it?

Communication requires cooperation. There is no point in talking unless someone is listening, and there is no point in listening unless it might be to the listener's advantage to do so. These two conditions are met only if there is a common interest between the parties. What can be the common interest between a raptor and a babbler?

A raptor can catch a babbler only by surprise, or when the babbler is far from cover. Babblers are masters of the thickets. Inside the vegetation, their strong feet, long tails, and short wings enable them to dodge and outmaneuver any predator. It is not unusual to see a group of babblers hopping and calling loudly in a bush or a tree that a raptor has perched in.

Indeed, at every loud noise, suspicious movement, or urgent warning call, babblers hurry into a thicket, where a predator has no chance against them; a predator that knows it has been seen by babblers is wasting its time if it stays in the area. And as long as the predator remains in the vicinity, the babblers have to stay near bushes and track the raptor's movements to make sure it does not catch them by surprise. If the babblers notify the raptor that they have seen it, both parties gain. The raptor moves on to another feeding ground, to try and surprise other prey; the babblers can resume their feeding. It makes sense for the babblers to signal to the raptor, and for the raptor to pay attention to their signal.

But one could argue that once the raptors have learned—or evolved the ability—to pay attention to the "warning calls," babblers could "cheat," emitting warning barks periodically, whether or not they see a raptor. The raptor, however, has a defense against such trickery: it puts the burden of proof on the babblers, reacting to calls only when it is sure that they are directed at it individually.

The behavior of the "warning" babbler fits this hypothesis: the babbler climbs to the top of the tree to bark, even though it could have watched the raptor and called from within the canopy—as it does when it is really frightened. The bird barks, disclosing its location, rather than trilling, as babblers do under other circumstances, and which makes them much more difficult to locate.[4] A babbler who would cheat by going to the top of the canopy and barking before it actually saw a predator would expose itself to raptors it might not have noticed. That risk helps ensure that if a babbler goes to the top of the tree and declares it has seen a raptor, it has indeed seen one.

If the raptor lands, the babblers often approach it and mob it, erasing any doubt the raptor may have that it is indeed the object of their calls. The calls change from barks to trills, interrupted by *tzwicks* whenever the raptor moves, which proves that the babblers are in fact watching it continuously. Bulbuls, blackstarts, sunbirds, warblers, and wheatears all join the gathering, so the raptor can see that they too are aware of its presence. By taking the calculated risk of approaching the predator, the birds increase the reliability of the message, which is that the predator has no chance of catching them. Indeed, they may even expose the raptor to its own enemies, such as bigger predators or humans, who may notice the commotion and come to investigate.

Having lost the chance to catch its prey by surprise and itself exposed to risk, the raptor is persuaded to leave the area. But mobbing exposes the babblers as well: Sordal[5] collected evidence showing that mobbing birds are occasionally caught by the raptors they mob. This risk is the price birds have to pay to convince the raptor that they are indeed aware of its presence.[6]

Unfortunately we cannot collect data on whether the behavior of the babblers encourages raptors to move on. Although the babblers are used to us and do not mind our moving among them, the raptors certainly do mind, and our presence causes them to move away.

STOTTING AS COMMUNICATION WITH PREDATORS

In the introduction we described the behavior of gazelles toward wolves. These behaviors were once thought to be warning signals meant to alert other gazelles.[7] But this explanation poses the same sort of questions as the contention that the babblers' barks are warning calls: the thumping, barking, and stotting—jumping straight up on all four legs—do indeed warn other gazelles of the predator, but why haven't gazelles evolved a more discreet way to alert each other? Again, we propose that the gazelle's behavior is in fact aimed at the predator. First, the gazelle shows the predator that it has seen it, by stamping its feet and turning the black-and-white "flag" of its behind towards the predator. Then it shows off its strength and fitness by jumping straight up. Only a gazelle certain of its ability to outrun a predator dares squander its strength in this way. Stotting provides a distant observer with clear evidence of the ability of the gazelle to jump: a high jump can be assessed from any direction, while long jumps could be properly evaluated only when observed from the side.

If the wolf wants to select a gazelle it has a chance to overtake, it makes sense for the predator to watch the stotting of gazelles before it chooses and starts pursuing its prey. The chase is a matter of life and death for the gazelle; it also requires a high investment from the wolf. Anybody who has seen a wolf after a chase, successful or otherwise, can testify to the wolf's exhaustion. If it fails, it may be a while before it can gather the strength for another attempt. Hence both the gazelle and the wolf benefit from communicating, as long as the communication is guaranteed to be reliable. The wolf avoids exhausting itself in pursuit of a gazelle it cannot catch and saves its energy for a more promising chase; likewise the gazelle will have the strength to flee from a predator who may feel up to the task of overtaking it.

Since 1977, when we first published the idea that stotting and "warning" vocalizations are in fact directed at predators,[8] zoologists have collected some evidence to support the hypothesis that individual prey benefit by signaling to predators. Fitzgibbon and Fanshawe[9] collected data in the 1980s in the  Serengeti to test this hypothesis. From a hilltop, they watched the hunting strategies of spotted hyenas and of hunting dogs, who catch their prey by running it down in a lengthy chase, rather than hunting by surprise, like cheetahs and lions. Some gazelles stotted repeatedly when dogs or hyenas approached, while others did not attempt to stot but rather fled right away. Both spotted hyenas and hunting dogs went after gazelles that did not stot, or stotted only a little; they avoided chasing those who gave an impressive display of stotting before their escape. Thus it seems that predators do indeed observe stotting in order to estimate an individual's ability to flee.

Hasson and his colleagues[10] collected evidence of similar behavior in zebra-tailed lizards. When a lizard is surprised in the open it runs away at top speed. But when it is near its hole or near a bush where it is able to hide, it does not flee but instead stops and moves its conspicuous tail from side to side. The model they used to explain the behavior of the lizard was the one we used to explain the behavior of the babblers and the gazelles.[11]

We were not the first to suggest that potential prey may communicate with predators. Smythe[12] observed the behavior of maras—large Patagonian rodents—when they meet humans. He realized that their behavior was analogous to that of stotting gazelles; but maras are solitary, and so he could not interpret that behavior as a warning to other members of their species. Smythe concluded that the fleeing maras signal to predators—that the former manipulate the latter, "inviting" pursuit only when they are ready for it.

Smythe believed that predators were stimulated to pursue fleeing prey even when they stood to lose by doing so. It did not occur to Smythe that prey and predators may have a common interest. He was aware of the shortcomings of his explanation; he suggested that evolution was not yet complete, and that eventually the predators might evolve the ability to ignore such temptations. But *any* feature whose benefits we do not understand can be dismissed by saying it is not yet well adapted and will change in the future. We think it more likely that most features *are* well adapted and in equilibrium with their environment; experience has shown us that this approach produces better explanations. Even when one has no ready explanation for a given phenomenon, it is best to assume that we are still missing a part of the picture and continue the search. Such a strategy is more likely to generate new findings.

The social interactions among children in a game of tag serve as a beautiful model of prey–predator interactions. Once a catcher—"it"—is chosen, children who know themselves to be slower than the catcher try to run as far away as possible. By contrast, those who know themselves to be faster than the catcher stay nearby or even approach the catcher, "inviting" pursuit. The catcher often ignores the stronger and pursues the weak, who seek refuge at the far end of the playground. The handicap—the risk of approaching the catcher—taken on by those who are confident in their ability to outrun him or her convinces the catcher not to waste any effort trying to catch them.

CALLS BY PREY DURING PURSUIT: THE MERLIN AND THE SKYLARK

Rhisiart, and later Cresswell,[13] studied how trained merlins—a species of falcon—select their prey among skylarks. When the falconer sights prey, he removes the merlin's hood and releases it to the chase. Rhisiart found that when the skylark sings while fleeing, the merlin is likely to abort the chase. When the lark does not sing, the merlin is more likely to continue the pursuit and is often able to catch the lark.

What could be the connection between the song and the chase? If we assume that some larks are faster than the merlin and some are slower, it makes sense for the merlin to try to select and chase individuals it can overtake. It is also in the interest of a skylark that flies faster than the merlin to let the merlin know that it cannot be caught. To convince the merlin of its superior abilities, the skylark must do something that a slower lark would not be able to do. Singing while flying is a good indicator of the lark's abilities, since it displays the bird's capacity to divert a part of its respiratory potential while still flying at least as fast as the merlin. A skylark that needs every ounce of strength it has to fly cannot sing at the same time.

We can see the similarity between the vocalizing of the chased skylark and that of children. A child fleeing during a play-chase often jeers at his pursuer. Such vocalization provides reliable information about the reserves of strength at the child's disposal. It makes sense to invest some air in a jeer if it saves one a greater investment in a prolonged flight. A child running as fast as he can cannot afford to jeer at his pursuer; if he does, the breathless quavering of his voice reveals that his energy will soon give out. The difficulty of vocalizing while running thus renders the jeering a reliable signal; it therefore makes sense for the pursuer to listen to vocalizations in order to assess the ability of the runner to flee. Both the pursuer and the pursued gain from communicating.

Once again we have drawn a parallel between the behavior of animals and that of humans. Some deride such comparisons as "anthropomorphisms" and consider them "unscientific." But a model is a tool; and anthropomorphic models, if anything, are closer to the reality of animal behavior than mathematical ones. Needless to say, models are not proofs; they are suggestions. Models can help one devise experiments and plan the collection of data to test interpretations. We use humans as a model only to improve our understanding of the behavior of other organisms and the logic behind their behavior. After all, mathematical models—much in vogue among zoologists these days—are also just that: models, which have to be tested, and which may or may not reflect reality.

WARNING COLORS (APOSEMATIC COLORATION)

Many poisonous animals have bright coloration that stands out from their surroundings. These bright colors—aposematic coloration—have long been considered a warning to predators, and indeed, many predators avoid these animals. But how did the coloring evolve? We suggest that the evolution of aposematic coloration parallels the evolution of all other signals that serve the common interest of prey and predator. To indicate reliably that the signaler is unpalatable, color must be so conspicuous that no *palatable* prey can afford to have it.

In fact, the showy colors *do* attract the notice of predators. There is now some evidence that naive and inexperienced predators actually learn to avoid these colorful animals by catching and tasting them;[14] although the predators may wound their gaudy prey while getting this first taste, they find it unpalatable and let it go. Indeed, many poisonous insects have especially strong cuticle and regenerative ability, which lets them recover from such injuries. Thus, by being highly colored the prey undertakes an increased risk of being spotted by predators, but that risk is balanced by the likelihood that most predators will avoid it.

What about mimics who look like poisonous animals but in fact are not poisonous? Some researchers are rather too prone to assume that a certain animal is mimicking another; one of the classic textbook examples of butterfly mimicry turned out to be even more toxic than the species it was "mimicking."[15]

Nevertheless, true mimicry does exist. There is a strictly limited evolutionary niche for such cheating: in cases where the receiver of the signal—the predator—does not benefit by making the effort needed to tell the difference between the poisonous species and the mimic. If the poison is very dangerous and the predator's food plentiful, and if there are few mimics, then the risk of its mistakenly eating the poisonous prey may outweigh the small benefit it gains by eating a mimic. But the greater the number of mimics and the scarcer the predator's food, the more tempting the cheaters become, and the greater the risk to them that a predator will learn to distinguish between them and their poisonous model. At that point, the mimics' prominent coloring becomes a severe drawback.

SIGNALING TO PREY

Not all communication between prey and predator runs from the former to the latter. The interaction between the tiger and the bull convinced us that predators may signal their prey as well. The tiger, once there is no chance it can catch the bull by surprise, moves in an arc around it. The bull endeavors constantly to face the tiger and point its horns at it, to prevent the tiger from pouncing on its back. The tiger, indeed, is looking for an opening to do just that, while trying to avoid the bull's sharp horns. The tiger's black and white ears, which point forward, make it easy to tell exactly what the tiger is looking at and so telegraph its intentions. Thus the tiger forces the bull to react; the tiger can then better assess the agility of the bull, its ability to defend itself, and its weaknesses. This information helps the tiger attack when and where it is sure to have an advantage over the bull, without running into its horns.

The tiger is telling the truth. It makes real preparations for attack and stops if and when it is clear that the bull is ready to repel the attack. The tiger wants to find a weakness in the bull's defenses; the more precisely the predator displays its location and its intended moves, the more reliable and informative the bull's reaction is, and the easier it is for the tiger to identify that weakness. If the tiger did not display its position and movements clearly, it could not rely on its prey's responses and might then underestimate the bull's agility, strike, and find itself impaled on the bull's horns.

The interaction between tiger and bull is similar to that between two boxers in a ring. They do not as a rule go directly for a knockout blow but first attempt exploratory punches. A major blow that misses its target would throw the attacker off balance and leave him vulnerable. Boxers usually start with blows that are strong enough to force the opponent to defend himself but do not involve too great a risk. The exploratory punches test the ability of the opponent to react. A knockout is usually attempted only after the attacker has assessed his rival.

Some see these exploratory punches as efforts to mislead the opponent. But misleading is a less effective strategy than one that utilizes real advantages, since its success depends on the opponent's stupidity and on chance, rather than on the attacker's real ability. Early in a bout, many of a boxer's moves are made to determine the opponent's agility in defense. We believe these exercises help the fighter—or the predator—not by misleading but rather by offering precise information about the attacker's repertoire. The more reliable the information provided by the attacker, the more he finds out; eventually, he can attack what he has learned are his opponent's most vulnerable points. The defender has no choice but to show the attacker something of his defensive ability. Of course, a defender certain of his strength and prowess may choose to ignore the attacker's maneuvers, showing a calm and a lack of response which themselves may persuade the aggressor of the defender's ability to withstand an attack.

Eshel[16] also suggested that predators may communicate with their prey in order to identify the most vulnerable individuals. The predator lets the prey know of its presence; at that point, differences in the behavior of the potential prey help the predator discern which are the most vulnerable. Kruuk[17] describes how spotted hyenas choose prey among a herd of gnu: one or several hyenas rush into the herd, then stop and survey the fleeing prey. Only when they find an individual in worse shape than the others do they—and any other hyenas who may be watching from the side—give chase. Even after such selection, only 30 to 40 percent of such chases are successful. The prohibitive odds against the hyenas' catching a physically fit gnu make crystal clear their interest in communicating with their prey.

The distance from which the predator "announces" itself must not be too great, though: otherwise all the prey animals will be able to escape. Eshel suggested that the colors of leopards and other spotted cats reflect this strategic fact. The spots on the leopard's coat merge into a uniform gray-brown, camouflaging the leopard, as long as the animal is too far away from its prey to pursue it effectively. But the same spots become distinct at closer range, making the leopard stand out and announcing the terrible menace to its prey, which then panic, showing the leopard which individual it ought to pursue. Lions, which are larger and hunt in groups, seem to choose individual prey less selectively. Their coloration, which is equally cryptic at all distances, enables them to come very close and take prey by surprise.

COOPERATION BETWEEN PREY AND PREDATORS WITHOUT COMMUNICATION

A common interest between prey and predator can evolve behaviors that serve that interest without communication. A flock of starlings fleeing a raptor often engage in very complex maneuvers: sharp turns, quick shifts of direction, re-

peated changes of altitude, and so on. This curious flight pattern has commonly been thought to be a communal defense system: the erratic turnings of the flock were believed to make it more difficult for the raptor to target a particular starling without colliding with another. But Eshel[18] suggested an additional explanation: that the maneuvers, which after all are physically strenuous, may be a strategy by which the starlings force the weakest of their number to fall behind as quickly as possible, to become easy prey for the raptor. That serves the interest of all other members of the flock, who thereby avoid a long, exhausting chase.

PRECONDITIONS FOR PREY–PREDATOR COMMUNICATION

Genuine warning behaviors do exist, of course. Since successful adaptations are those that increase successful reproduction, parents may benefit by warning their offspring, even when they put themselves at risk. Also, as we will see in chapter 6, the babbler who emits the so-called "warning calls" may increase its prestige within its group. But again and again we find that behaviors that may seem to serve as warnings to other members of the species, and that have thus been considered altruistic, are in fact better understood to be signals sent to predators to serve the interest of the individual sending them.

Communication, again, requires collaboration. In order for communication to take place, both the signaler and the receiver must benefit from it. Two conditions are necessary for communication to evolve, and they are especially clear in the case of prey–predator interaction: the parties must have a common interest, and the signals used must be ones that cannot be faked.

Communication between prey and predator makes sense whenever some of the prey can escape the predator and the latter has to target one of the laggards. Under these conditions, a mechanism for such communication will probably evolve. This is not necessarily a tribute to the wisdom of either predator or prey. Rather, it is testimony to the power of natural selection to evolve in prey behaviors that will help them escape, and in predators the ability to heed signals that will spare them fruitless chases. Such systems may evolve even in microorganisms—between bacteria and their hosts, as we shall see in chapter 16.

All animals, humans included, can behave wisely without being aware of the wisdom of their actions or the logic behind them. This wisdom and this logic are expressed in a behavioral mechanism that evolved through natural selection, a process by which those who adopt effective mechanisms survive, while those who do not perish, or do not reproduce to the same degree. Both predator and prey are probably unaware of the significance of their behaviors, just as they are unaware of the functioning of their brains, their kidneys, or their muscles. There is no reason to admire these behaviors any more than we admire the functioning of other biological mechanisms—or any less. All are marvels that deserve our appreciation.

CHAPTER 2

COMMUNICATION BETWEEN RIVALS

THREATS AS A SUBSTITUTE FOR AGGRESSION: ARE HUMANS WORSE THAN BEASTS?

Rivals rarely attack each other without first signaling their intentions. Most of the time, they do not attack at all, and the conflict between them is resolved by an exchange of threats. These signals come in many forms: singing, aerial displays, electric pulses (in the electric eel), puffs of noxious chemicals, posturing. The song of the nightingale advertises the bird's readiness to defend its territory and deters its rivals. Red deer walk side by side and roar. Fish swim in parallel to one another, extending their fins.

What about humans? There is a common notion, endorsed by Konrad Lorenz, that the tendency to escalate conflicts into combat and the readiness to kill opponents is unique among humans.[1] But in fact the typical study of animals done at the time simply did not last long enough for researchers to determine this. In most long-term studies observers have recorded conflicts that escalate into fights, which have occasionally resulted in woundings and even death. In our own long-term study of the babblers at Hatzeva—over 20,000 person-hours of observation—there have been eyewitness accounts of some 20 fights that resulted in killings; we have found indirect evidence that several more have occurred.

Among humans, too, most conflicts are resolved by threats rather than by actual violence. This is true in particular on the personal level, but also on the interna-

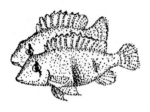

tional level. Wars are terrible, and partly for that very reason, most conflicts are resolved without them. The reason we regard humans as uniquely aggressive is the degree of attention we and our news media pay to cases that do escalate into violence. If we were to turn off the TV, cancel the newspaper, and rely on our own observations—as does a zoologist studying an animal species—the picture would be very different. We would conclude, rightly, that humans do not as a rule kill or wound each other. Humans, like other animals, resolve most conflicts by communicating—which often includes the exchange of threats.

Mammals, birds, reptiles, fish, insects—in fact, all living creatures that communicate in any way—use threats. Resolving a conflict by threats alone prevents the loss of time and energy, and the risk of injury or death, that attends an actual battle. It is easy to understand what the winner gains by using threats rather than fighting; but why should threats make the other party back down? What is it that convinces one of the opponents to give up a bit of food, a potential mate, a territory without even trying to fight? Maynard Smith and Parker[2] put forward the obvious argument that if one is going to lose anyway, it is better to lose without being defeated in a fight. Yet how does one know one is going to lose? What is it that convinces one of the two rivals that its defeat is inevitable, or that what it would gain by winning is not worth the cost it would incur by fighting?

HANDICAPS AS KEYS FOR RELIABLE THREATS

In 1977 we suggested a solution:[3] the threats themselves can be reliable indicators of each rival's chances in a fight. Reliable threats, by definition, are signals that enable one rival, the one who is likelier to win a fight, to threaten more effectively than the one who is likelier to lose. How can a signal work this way? The threat itself must increase the risk that the threatener will be attacked, or will be at a disadvantage if attacked; an animal that is genuinely willing to fight and confident of its ability will find such a risk reasonable, but one lacking strength or motivation will find the risk excessive and thus be unable to threaten to the same degree. In other words, for a threat to be reliable, the signal must increase the danger to the threatener—and an escalation of the threat must increase that danger even further.

There has to be a fundamental relationship between the particular pattern, or form, of a reliable signal and the specific message the signal conveys—to be more exact, between the cost of that pattern and the message.

What is an honest threat? An honest threat communicates reliably one's ability and willingness to fight. A reliable threat leaves the signaler open to attack. This increased risk is acceptable to the honest signaler—the one who thinks the objective is worth a fight, and that it can win against that particular opponent. Such a threatener has already decided to fight if its opponent does not retreat, and the increased likelihood that it will actually have to fight does not deter it. A bluffer— one who tries to gain by threats alone but is not really willing to back up its threats by fighting—would find the increased likelihood of being attacked too risky. A quick look at known threat signals illustrates this point.

THREATS BY APPROACHING

Threatening by approaching a rival is very common indeed. Approaching is a reliable threat because by approaching its rival, the signaler opens itself to attack. If such signals were a pure convention, then any movement would be as likely as any other to take on the meaning of threat. If the most clearly visible movement were the aim, then a movement sideways would have been better than a movement forward or backward. But a movement sideways is less risky and thus, though clearer, is less reliable. Likewise, if signals were arbitrary conventions, a movement backward would have been just as likely to become a threat as a movement forward. We know of no case, however, in which a movement away from one's rival has become such a signal.

The stance of the defiant human male is well known to us all: a straight back, the chest thrown out, the shoulders back, chin up. This is a very inefficient and risky posture in which to enter hand-to-hand combat. The uptilted chin is exposed to blows; the erect body makes it difficult for the threatener to launch a surprise attack or to change position at all. It is the exact opposite of a boxer's or wrestler's stance in the ring: ready to attack or avoid an attack, the boxer keeps his chin down, close to his chest, and his body coiled like  a spring; he balances on the balls of his feet, on the alert, ready to seize the first opportunity. Of course, boxers cannot resolve the match by threats: both have already committed themselves to fight.

The threatener, by contrast, uses precisely this vulnerability to strengthen his threat. By standing up straight he gives up the benefit of a good defensive stance and the option of a surprise attack. In the days before razors, a man's thrown-out chin presented another risk: it brought the threatener's beard nearer to his rival and made it easier for the latter to grab it. By putting his chin out, a threatener shows his confidence that his rival will not dare or will not be able to grab him by the beard or punch him on the chin—and that he, the threatener, is still confident of winning the fight if the other does dare. A version of this signal is familiar from a hundred movies and cartoons: the tough guy points to his outthrust chin and taunts, "C'mon, big shot, lay one on me!"

THREATS BY STRETCHING—DECEIT OR EXPOSURE TO DANGER?

Many animals threaten by stretching their bodies. In such cases, the threateners do not show their weapons—teeth or claws—to their opponents but rather expose their whole bodies to attack. Threatening dogs, a familiar example, stand side by side on tiptoe, bodies stretched and hair raised.[4] Such a posture is commonly explained as an attempt by the threatener to present itself as bigger than it really is and thus to deter rivals.[5] The stretched body does indeed provide precise information about the threatener's size, but stretch as it may, no animal can become bigger than it really is, and the smaller of the two will still seem smaller than the other.

Stretching is therefore not likely to make a rival misjudge one's size. Is it done, then, to show off one's *actual* size? Apparently not: even rivals who know each other well, such as longtime neighbors who are well aware of each other's true size, threaten by stretching their bodies. The only point in signaling is to affect the other's behavior, and a fact that is known already is unlikely to change that behavior. Keeping its body stretched exposes the threatener to risk; why take a risk in order to convey information that the rival has already?

The point of the stretched body is not to show size but to convey confidence. Key to any prediction about the results of a fight is the willingness of a given individual to invest in that fight—in other words, its motivation. One who is not sure the object of the fight is worth the risk of injury will hesitate before exposing its body to danger. Motivation changes with circumstances—unlike body size, which is easily determined and does not change from day to day. The appearance of another rival on the scene, the approach of a predator, the sound of one's offspring calling for help—all can instantly change one's willingness to take risks. Thus, stretching one's body in the presence of a rival is a reliable, moment-by-moment indicator of one's current willingness to engage in the prospective conflict. Such stretching displays often last for a good while before coming to a resolution: Clutton-Brock and his colleagues describe red deer walking in parallel along their territorial line, bodies stretched, for half an hour before one of them decides to attack or to withdraw.[6]

Like stretching, bristling hair among mammals and the stretched fins of threatening fish are commonly explained as attempts to appear bigger. Such explanations assume that the rival cannot detect the truth. Indeed, fish with stretched fins look much bigger than fish that hold their fins closely folded; but since the strength of a fish in a fight depends on its muscles, not its fins, this deceit would only work if the rival mistakes the fins for muscles. Yet the coloring of the fins makes it unlikely that any such deceit is intended: the pattern and the color of the fin are almost always different from those of the body, and the stretched fins thus look very different from the fish's muscular body.[7]

A threatening fish stretching its fins is both revealing the precise size of its muscles and exposing its body to attack. By contrast, a fish that is having the worst of a battle but is still willing to fight finds refuge in a corner of the aquarium or in a narrow gap between rocks, fins held tight and ready for instant motion, teeth exposed. It is ready to bite its attacker at the first opportunity and is minimizing its rival's ability to get at it. The fish that is losing the battle cannot afford to expend much on showing off; it conveys only its grim willingness to go on fighting. Its position is quite unlike the threatening stance of a fish confident that it can win, which shows off its self-assurance with a stretched posture that prevents it from springing into action and opens it up to its rival's attack.

THREATS BY VOCALIZATION

Morton[8] noticed that in vertebrates, threatening vocalizations are usually low in frequency, in contrast to appeasement calls, which are more high-pitched. Morton also observed that larger individuals usually have lower voices than smaller ones and suggested that a low-frequency call is intended to reflect the threatener's size.

This raises a number of questions: When rivals are looking at each other, why should they use hearing rather than sight to assess each other's size? If a low voice can frighten rivals, how is it that animals have not simply evolved longer vocal cords, so as to threaten better? And most of all, during conflict, why does the same individual emit high-pitched sounds when frightened rather than sticking to the lowest-pitched sounds it can make?

The pitch of the voice reliably discloses the tension of the signaler's body.[9] A tense body makes a more high-pitched sound than a relaxed one. A frightened individual is tensed to take flight or to fight back. Only one who is relaxed, not poised to take instant action, can sound a low-pitched, threatening note. Such an individual discloses reliably that it does not fear its rival; it is not coiled like a tightly wound spring and thus has exposed itself to a first strike. This—the cost involved in making such a sound in a rival's presence—is the very element that makes the message reliable. The threatening calls of two rivals confronting each other are true indicators of each one's willingness to risk attack.

In a famous experiment, Davies and Halliday[10] found a correlation between the pitch of male toads' croaks and their ability to fend off challengers in the contest over a female. A male toad finds a female ready to lay and grips (amplexes) her, waiting for her to discharge her eggs so that he can fertilize them. Rival toads try to kick him off and take his place. In response, the first croaks. Davies and Halliday found that the croaks of the amplexing toad correlated with its body size, and that larger toads displaced smaller ones. They suggested that the pitch of the call displayed the size of the threatening toad and thereby its potential to win the fight.

But why should other toads listen to the pitch of its croaks to assess the size of a toad that is right in front of them? We suggest that the croak displays not size but rather the toad's self-confidence. If a toad riding a female with his rival present fears being kicked off, he grabs her tightly—rendering himself unable to emit a low, relaxed croak. Only one certain of his ability to stay on no matter what, or confident that his rival will not dare challenge him, can afford to relax enough to emit a low-pitched croak. Obviously, this confidence has something to do with body size: smaller toads would naturally fear larger ones. But if pitch indicates only body size, each toad's croaks would stay at the same pitch no matter which rival it was confronting; if pitch reflects confidence and motivation rather than size, though, then the same toad would emit a low, relaxed croak when confronting a small rival and a higher, tenser one when confronting one larger than himself. It would be simple and interesting to test this hypothesis.

Among humans, too, posture affects the ability to produce a persuasive vocal threat. Actors can sound convincingly threatening even though they have no intention at all of actually fighting with an onstage "rival." We were told by Nissan Nativ, acting teacher and head of a theater school in Tel Aviv, that an actor does not have to worry about how to sound menacing: once he or she assumes the body posture of a confident, aggressive individual, the correct tonal quality comes of itself.

Does this mean that actors would make perfect bluffers in the real world? Not at all. The actor is not actually in danger on stage. In real life, facing an actual enemy who might attack at any moment, even the best actor would find it just about impossible to bluff and threaten with a relaxed, low-pitched voice. Any tension in his body—any readiness to fight or flee instantly—would reflect itself in his vocalization.

Vocal threats are also used from a distance, when attack is not yet imminent and tension versus relaxation is irrelevant. In such cases the vocal signal conveys other qualities. One of them we have learned, as we have so many things, from the babblers.[11] The long-distance threat call of babblers is a sequence of several loud syllables clearly separated by precise intervals. It gives a good indication of the motivation of the caller and its willingness to fight. When a group of babblers advances toward a border clash with another group, one can easily distinguish the calls of the front-runners from the calls of those lagging behind. The ones in the front emit rhythmic and clearly defined syllables; those in the rear, who seem less eager to fight, emit softer sounds, with less precise intervals. Once the rival groups engage, the calls of those actually fighting become metallic and higher in pitch, and the intervals between the syllables are less precise. One can "hear" the tension in their bodies.

Why do the babblers use precisely spaced syllables only when they are eager to fight? In order to emit rhythmic, regularly spaced, and clearly defined syllables, one has to concentrate on the act of calling. Any distraction—such as a glance sideways—distorts both the rhythm and the precision of sound; an individual cannot at one and the same time collect information and concentrate on performance. A call composed of precise, rhythmic syllables testifies that the caller is deliberately depriving itself of information, which means either that it is very sure of itself or that it is very motivated to attack, or both.[12] A human being in control of a situation, too, tends to issue threats in an ordered, rhythmic cadence. The even beat increases the effectiveness of the threat, since it shows confidence and demonstrates the threatener's ability to concentrate on the threat regardless of the fact that he or she may thus be deprived of crucial information.

OTHER THREATS

Staring at a rival also impairs one's ability to collect information; the risk is small for a dominant individual, who is not likely to be attacked from behind in its own territory or group, but it may be too high for one less in control of its surroundings. Hence, staring is perceived as a threat. The individual that stares is dangerous either because it is dominant within its group or because it is so highly motivated to attack that it does not seek to gather all the information it can.

Contempt can also be used to deter rivals. A Hollywood sheriff entering a rustler's hideout with arms folded and pistols holstered is showing his confidence: if he were to enter gun in hand, it might increase the chances of an immediate victory, but by displaying his confidence, he may get his man without the need for any gunplay. Likewise one who turns his back on a rival shows his contempt—and his confidence; a dog turning aside to urinate in the middle of a confrontation shows it is confident that his opponent will not dare attack him—and that if attacked, he will still win, even when caught with one leg in the air.

Still another way to issue a threat is to show the degree of harm one is willing to suffer in order to win. When ants use formic acid and bees use chemicals in their fighting, they create an environment hostile to both signaler and rival. It makes sense that one able and willing to endure the noxious chemicals displays in a reliable way its ability and commitment to the fight.

SOCIAL HIERARCHIES AND DUELS BETWEEN EQUALS

Most conflicts, in fact, occur between rivals that have clashed with each other in the past. In such cases the previous loser tends to submit without fighting. In social systems where the same individuals meet repeatedly, these histories of previous conflict produce a social hierarchy—a pecking order—in which each individual knows its rank from experience and submits to higher-ranking individuals. Often the mere hint of a threat by a dominant individual is enough to deter a subordinate.[13]

When both rivals are the same size and equally strong and willing to fight, one rival cannot threaten reliably more than the other, and there is no way to win without fighting. Darling,[14] who studied red deer, mentions that actual physical clashes occur only between males of similar size; in other cases, one of the contestants withdraws following an exchange of threats. Barrete and Vandal[15] observed 1314 confrontations between reindeer. Only in six of these did the exchange of threats develop into an actual fight; in all six cases, the males were very evenly matched in size and strength. But even between rivals who are equally strong and motivated, threats still have their use: any injury to one of the contestants in the fight changes the balance of strength, opening the way to another round of exchanged threats, which may well persuade one that he no longer has a chance of winning and should withdraw.

• • •

CAN THREAT SIGNALS EVOLVE FOR THE GOOD OF THE GROUP? THE DRAWBACKS OF "GROUP SELECTION"

According to the theory of "group selection," groups that use threats instead of aggression have an advantage over and eventually supplant groups whose members always fight each other to the bitter end.[16] There was a major discussion of this idea in the 1960s, which resulted in the practically unanimous conclusion that group selection cannot explain the evolution of traits, except perhaps under very special circumstances.[17] Yet explanations based on group selection are still common in popular literature.

According to group selection theory, a population—a group of individuals—that solves conflicts by threats rather than by fighting has the advantage over a population of aggressors who wound and kill each other. The efficient group supplants the less efficient, more self-destructive one. According to this theory, selection can act on the group as a unit—thus the name "group selection."[18] This model—which has not been supported by observation—suggests that each individual submits to threats from other group members because fighting would harm the group. By this logic, there is no need for any connection between a threat and the actual ability of the threatener. Any gesture can be a threat, as long as it is so accepted by the group.

The problem is that if such signals are only conventions and bear no connection to the threatener's real ability, then a mutant who does not obey the "rules"—who fights rather than giving up something important just because it is threatened—will reproduce better than one who submits. Its descendants will inherit the trait of fighting rather than obeying the rules, and they too will reproduce better, until they form a significant part of the group. Thus over time the group will lose the shared advantage of using threats rather than fighting.

Some argue that society will punish those who break the rules. This only expands the problem, which becomes: Who is this "society" that will impose punishment? How will the punishing work? The exception, of course, is when the individuals meting out punishment receive direct compensation for their efforts, like police and judges in human society.[19]

A group-selectionist might retort that a group in which combative dissidents come to prevail will ultimately die out because of excessive aggression. But before that happens, what is to stop some of the rule-breakers from moving into a "non-aggressive" group of their species—where they would again reproduce more successfully than other group members until that group, too, loses the ability to avoid violence by using threats? A trait that harms or deprives the individual who bears it and only benefits "the group" cannot survive natural selection, even if threatening rather than fighting is "good for the group."

In contrast, our model, the Handicap Principle, shows how the pattern of the signal itself can prevent cheating and can show the real motivation and ability of the threatener in a reliable way. The patterns of threat signals—like those of other signals—are neither random nor arbitrary. The need for reliability links the signal directly to the message it conveys and guarantees that the cost of the signal is reasonable for an honest signaler but prohibitive for a cheater. In the case of a threat, the cost that the honest signaler is willing to undertake, and that a cheater cannot afford, is the risk that the signal entails. Rather than being a mere convention, the signal itself provides the genuine information that is needed to resolve the conflict. Thus there is an optimal signal for every threat—and indeed for every particular message. The Handicap Principle not only asserts that this is so but explains why.

CHAPTER 3

MATE SELECTION

Sexual displays often attain gigantic dimensions and take bizarre forms. The peacock's tail grows into the well-known large, colorful fan; peacocks presenting themselves to peahens spend time and energy holding their tails spread open and upright and vibrating them rapidly. Males of many other bird species grow long tail feathers for sexual displays. Relative to their size, some male pheasants and birds of paradise have tails almost as large as the peacock's, and the males of some species of songbird, such as the viduas of Africa, have tails that are proportionately even longer.

Male songbirds sing to attract mates; some of them invest most of the day in singing. Others, including many larks, sing during strenuous flight displays. Male sage grouse and male ruffs dance in special arenas for many days to compete for the favors of females. Blackcocks vocalize and display their tails as they dance, broadcasting their message by means of several modalities at once.

Sexual display is not unique to birds. Mammals, reptiles, amphibians, fish, insects—all invest in sexual displays, and in each of these groups there are species in which the investment attains striking, even fantastic dimensions. Crickets, grasshoppers, and cicadas sing for many hours to attract partners. Fireflies display with flashes of light. Moths and many mammals emit scents. Even one-called organisms like yeast and algae emit pheromones to attract mates.

Sometimes bodily growths are means of sexual display: newts grow a finlike nuptial crest along their backs; the white pelican grows a bump of flesh between its eyes. The males of many bird species such as shrikes, terns, and gulls, as well as some insects, court their mates with nuptial gifts of food, leaves, or twigs. Male bowerbirds build complex bowers and decorate them with shells and flowers, bones, insect skeletons, and colorful fruits to attract females; some male fish and crabs build sand castles on rocks exposed to the waves, castles that have to be continually rebuilt.

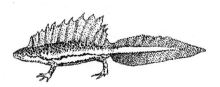

The extravagant dimensions of sexual displays make them seem like crazy fashion shows. But have they run beyond reasonable evolutionary control, as is commonly assumed? Or do the dimensions of the displays and the specific forms they take serve a purpose?

THE CONFLICT INHERENT IN COURTSHIP

Williams[1] emphasized the competitive aspect of courtship: males and females have conflicting interests. Each wants the highest-quality mate it can get—the mate that can best improve its offspring's genes and, depending on the gender and species, best raise those young. Williams therefore suggested that during courtship males and females can be seen as opponents. The male, like a good salesman, does whatever he can to impress females, while the goal of the female, like that of a shrewd customer, is to check the merchandise and accept only proven quality.

Of course, females also advertise themselves to males, with the same ensuing conflict of interests.[2] Still, the possibilities open to them are different. The number of a female's offspring is limited most by her own capacity to produce eggs or undergo pregnancies, while a male's breeding success depends more on the number as well as the quality of the females he can persuade to breed with him. For the sake of convenience, and since males as a rule invest more in advertising than females, we will discuss the issue mostly in terms of males as presenters and females as choosers.

How can males prove themselves to be superior? Williams did not address that question. But in the previous chapters we have seen how reliable communication develops between enemies of different species and between rivals within a species. The same logic applies to sexual display.[3] Here too the conflict of interests—between male and female—is often resolved by communication, which depends on the evolution of reliable signals.

Signals are reliable if a cheater cannot gain by using them—if the investment in the signal is a reasonable one for a truthful suitor to make, but prohibitive or unprofitable for a cheater. The more the suitor stands to gain, and the bigger the loss to one who accepts a false suitor, the more the signaler must invest in the signal in order to reliably demonstrate his superiority.

We assume that the specific investment a signaler makes is directly linked to the message of the signal. Male rivals are only interested in the fighting ability of their adversaries; predators are concerned only with their prey's ability to escape. Individuals looking for mates, however, are interested in a wide range of qualities.

What do animals look for in a mate? That depends. Where both male and female take care of the young, the ideal mate is not just of superior genetic quality—quality that will be passed on to the offspring—but also skilled enough to provide for its family and committed to parenting the offspring effectively. In such species, the male can commit to only one female at a time—or to at most a few. A female in such a species may well have to compromise on quality in order to get a male willing to commit to her. At the other end of the spectrum are species in which males have no parental involvement whatsoever. In these species a female can concentrate on finding the most superior sperm donor she can, even if she has to share his favors with many other females. In such species a few outstanding males get most of the females, while young and low-quality males do not copulate at all.

Courtship signals thus convey different messages in different species. As with all signals, we expect to find a direct relationship between the investment in sexual advertising and the specific information the courtship signal conveys to the selecting party.

COURTSHIP HANDICAPS AND THE INFORMATION THEY CONVEY

Feeding Ability

Courtship feeding—as in terns, shrikes, and great tits—reliably demonstrates both the feeder's ability to give up a good portion of the food he collects and his interest in the particular female. The female gains both the actual food provided and the knowledge that the male would be a good provider for her offspring. The more food the male brings to the female, the more reliable the message: that he is a good collector of food. Indeed, Nisbet[4] found that male terns who provide more food during courtship also feed their offspring better. Courtship feeding is a much better indicator of the male's quality as a hunter than a mere show of strength would be. The effort required to feed the female prevents pretense by a male who can barely feed himself. It also prevents males from courting many females simultaneously.

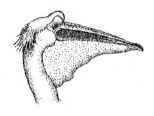

White pelicans, both male and female, grow fleshy bumps between their eyes when they are ready to breed. The bump interferes with the pelican's ability to see the area around the tip of its bill. In order to catch prey, a pelican with a bump has to remember where it last saw its prey and project the prey's likely movements. An inexperienced or inept pelican would not be able to do so. A pelican that can fish and maintain itself in spite of the handicap of its bump is reliably demonstrating its expertise in fishing. Later, when the pelicans have to feed their brood of four or five demanding young, the bump shrinks and they are able again to hunt more efficiently.

Singing can also demonstrate the ability to provide. The time invested in singing cannot be used for foraging. A courting male who handicaps himself by singing continuously provides evidence that he needs less time to forage, either because he is very efficient or because his territory is very rich. Wilhelm and his colleagues[5] studied the effect of supplementary feeding on the singing of yellow-bellied sunbirds. They found that males who were not given insects did not sing, while those who received insects and sugar water sang often and at length. Time spent in sentinel activity, or in dancing displays, can also indicate expertise in finding food, especially when the "waste" of time comes early in the morning after a long, cold night without food.

Superterritories

O'Donald[6] has suggested that the size of an animal's territory can serve as an advertisement. Males and females of many species protect a territory and chase others away from it. Often the territory is far larger than is needed to provide food or shelter for a pair of mates and their offspring. Some suggest that these bigger territories prevent overexploitation of food resources, and thus a richer resource remains to sustain the population as a whole in the future.[7] But this argument depends on the mechanism of group selection, which, as we have seen, is questionable. When some males protect territories that are larger than they need, the population as a whole might gain from the preservation of future food resources; but the males who hold smaller territories would gain the most, since they would enjoy the resources preserved by earlier residents without squandering their own efforts protecting larger territories than they need; they would therefore be able to devote more of their energy to reproduction. Thus, over the generations, the tendency to hold a larger territory than one needs would disappear. Why, then, should an individual spend time and energy and even fight to protect a bigger territory than it needs?

The rich build grand homes and mansions not because they need them as shelter for themselves or their children but in order to proclaim their status. The cost of building and maintaining a palatial mansion advertises the owner's wealth to his competitors and colleagues. Just so among many species of animals: a large territory proves the male's superiority and attracts a good mate. Indeed, observations show that females settle first in the larger and richer territories; very small territories may attract no females at all.

These superterritories may indeed preserve the resources available to the whole population of the species; but this is only a by-product and does not play any direct role in the evolution of the trait of establishing superterritories. A good analogy is provided by the vast hunting lands maintained by European noblemen in past centuries,  while throngs of peasants starved. These territories were used for sport hunting and were strictly guarded against poachers. Such superterritories served to show off their owners' wealth, authority, and power to their peers, whom they invited to the hunt. Many of these same hunting grounds by now have become national parks, which preserve for people of our own time the animals and the natural forests of Europe—but this is hardly the reason the noblemen established and maintained their hunting estates generations ago. In just the same way, males hold large territories because by doing so they deter their rivals and attract good females. Large territories may indeed preserve the species' resources of food against over-exploitation—but that is a side effect, not the superterritories' actual purpose.

Courtship Vocalization

Many animals vocalize during courtship. Lions, tigers, and deer roar; cicadas and crickets chirp; birds sing. Courtship calls can be dangerous: Ryan and his colleagues[8] showed that frog-eating bats locate frogs by the amphibians' courtship calls. Only a male frog that can successfully avoid bats despite disclosing its location to them can afford to croak much. Courtship calls can also be very demanding: Clutton-Brock and Albon[9] found that red deer are often exhausted after a roaring contest with rivals. Only a strong, well-muscled individual can roar loudly for a long time.

The details of a call, its tempo, and the number of syllables in a phrase can demonstrate the caller's quality. The song of the great tit is a series of precisely spaced syllables. Lambrechts and Dhondt[10] found a positive correlation between the number of syllables in a phrase and the rhythmic precision of the last few syllables on the one hand, and the reproductive success of the singer on the other. The ability to maintain both the tempo and the pattern of syllables at the end of a long phrase would seem to be a good indicator of quality. As was discussed in chapter 2, the precise performance of a call demands concentration. A less able male would probably find it difficult to concentrate for long; after all, the song reveals his location, and he must keep an eye out for approaching rivals and predators. The singing conveys his confidence or lack of it, information important to a female who must decide whether to accept him as her mate.[11]

Colors

The adult males of many species of birds are far more colorful than females and young males: examples include peacocks, ducks, birds of paradise, and sunbirds. Colorful plumage attracts rivals and predators and thus serves as a reliable signal of quality: only males of high quality can risk advertising their location. Conspicuous coloration also emphasizes the exact shape, posture, and movements of its bearer. A high-quality individual "wears" bright coloring well; on a low-quality one the same coloring accentuates imperfections.[12]

Scents (Pheromones)

Scents also serve to attract mates. Studies have found that some female insects can identify a dominant male by its scent alone.[13] Many male mammals, too, use scent to attract females and deter rivals. We know a great deal about the chemistry of pheromones—chemicals that are produced by one individual in order to influence the actions of others; but very few scientists have tried to explain the adaptive significance of specific chemicals. It would be fascinating to discover what information the pheromones of each species provide about their producers.

The main component of the pheromone secreted by male arcteid and danaid butterflies is a derivative of an alkaloid—a strong poison produced by plants for their own protection.[14] Arcteid larvae can metabolize the poison and thus take advantage of a food source that is not available to most other animals; the poison that they take into their system then helps protect them from predators. A male arcteid butterfly secreting this pheromone testifies that as a larva he was able to feed on the poisonous plants; the concentration of poison in the pheromone demonstrates the male's relative physiological ability to deal with the poison. Danaid butterflies secrete a similar pheromone, derived from plants that they eat as adults. Eisner and Meinwald further proposed that the pheromone could function as a chemical yardstick by which females gauge the poison load their suitors carry. The alkaloid in the pheromone may also show that the male will probably be able to pass on a good quantity of alkaloid to the female during copulation, to protect her and their offspring against predators.

The monarch butterfly is one species in the danaid family that does not use these alkaloid derivatives in its courtship. It is a migratory species that winters every year in California, Mexico, and Florida. This species instead courts its females by means of aggressive displays. Males bounce on passing females and throw them to the ground, where they copulate.[15] The ability of the male to use force in copulation may attest to his strength and stamina, which are necessary for long migrations.

Artifacts and Constructions

A cichlid fish of Lake Malawi in Africa gathers a pile of sand and forms a depression in it for the female to lay eggs in. This nest is not used to raise young: both sand and eggs get washed away by the waves in the lake.[16] In our opinion, it is precisely the difficulty involved in gathering a pile of sand again and again in a wave-swept area that enables the male to show off his ability reliably. Crabs who live in the tidal area also build sand castles, which have to be rebuilt at every low tide.[17]

As mentioned earlier, courtship displays are especially elaborate in species where the males do not take care of offspring and are free to concentrate their efforts on attracting females all season long. The top performers get most of the females, while most males do not copulate at all. The most famous of these species include peacocks, ruffs, grouse, manakins, birds of paradise, and bowerbirds.

Male bowerbirds, a family found only in Australia and Papua, spend much of their time building bowers of twigs on the ground. These structures have no utilitarian purpose; their only function is to serve as stages on which males perform courtship displays. The bowers of each species have their own structure and characteristic decoration.[18] One species builds a platform of twigs a meter in diameter; on the platform, two parallel rows of standing twigs, each half a meter long, form a corridor in which copulation takes place. The male of another species builds a mossy wall about four inches high around a courtyard, at the center of which, as a rule, is a sapling; the bird covers a meter of the sapling's trunk with woven twigs. Still another species covers the courtyard, making a hut with only one small entrance. The builder adorns his bower with rare ornaments, such as the feathers of birds of paradise, or with fresh flowers, which must be continually replaced, showing off his ability to find decorations. In fact, the two highly decorative feathers on the head of the king-of-Saxony bird of paradise are valued both by bowerbirds and by tribesmen of Papua New Guinea.[19]

The female visits the bower several times and carefully examines both structure and builder before deciding whether to copulate with the hopeful candidate. The female copulates only once before egg-laying. She may well visit a bower, examine it, watch the male perform his dances, even take part in the dancing and court the male before deciding to pass him up. It seems that the decoration of the bower plays a large part in the female's choice of a father for her offspring.

The number of decorations is important to females too. They prefer males whose bowers are richly decorated. Males steal decorations from neighboring bowers. Borgia tried adding individually marked rare and sought-after ornaments to one set of bowers, then to another; he found that no matter which bowers he added the ornaments to, they always wound up in the bowers of the most successful males.[20] To be attractive to females, a male bowerbird has to spend considerable time and talent building his bower, collecting decorations, placing them, and guarding them against his neighbors, not to mention performing on the stage he has built. He thus proves that he is stronger and more energetic than his neighbors, that he can feed himself adequately and is still able to build and decorate his elaborate bower, guard it against competitors, and raid their bowers.[21]

When we visited Borgia in Australia, we saw some bowers of the great bowerbird. The platform at the entrance to the avenue included flat stones from a riverbed, bleached bones (mostly vertebrae), broken glass—most of it green—and colorful bits of foil and cardboard. Borgia observed that the arrangement of the decorations was not random. The green bits of glass were consistently placed by the northern entry, which in the southern hemisphere faces the sun. The glass glistens in the sunlight, and the male displays his violet plumes against a shiny color-coordinated backdrop.

Combinations of Signals

In most cases, several signals of different kinds are used in courtship. In birds, special feathers, bright colors, singing and calling, dancing, and gift-giving—the last three of which demand time and energy—all play their part. Each modality brings out a particular quality of the male; the female can then use several criteria

to assess the male. Let's take the display of the peacock as an example. The male holds his tail upright and spread out—which demands considerable effort. From time to time he shakes his tail vigorously; this requires yet more effort and produces a remarkable rattle. The "eye" patterns on the peacock's tail, the glisten of his feathers, the crown on his head, all add up to a symphony of shape, color, pattern, movement and sound—a performance that is announced with periodic roars.

Each aspect of the display seems to convey specific, reliable information about a particular feature of the male. The long tail feathers are grown over a period of several months, during a time of the year when food is scarce. Unhealthy birds arrest the process, so a male who displays a set of perfect tail feathers advertises that he has been in good health and has managed to find food even during molt season.

The long, heavy, brightly colored tail also attests to the owner's strength and skill, for he has succeeded in avoiding predators despite such a burden. By holding his tail upright and shaking it, the peacock proves his stamina, and his roars show he is not afraid to disclose his location to rivals and predators. The perfection of the tail's pattern testifies that the peacock's development while the tail grew was perfectly coordinated, as we shall see in chapter 8. Each of these criteria seems to be minutely scrutinized by females: Petrie and her colleagues found that taking out even five of the hundred and fifty or so feathers in a peacock's tail reduces his ability to attract females to his dancing arena.[22]

Long Tails

Moller[23] investigated the long outer tail feathers of barn swallows, small songbirds that catch their food in flight. Barn swallows have long, forked tails; the outer feathers are longer in adult males than in females or in young males. When Moller added extra pieces of tail feather to the tails of some males and shortened the tail feathers of others, he discovered that those with the longer tails, whether natural or artificially enhanced, found mates more easily than those with the shorter or shortened tails, and that they got to copulate with additional females as well (extrapair copulation). But the inadvertent cheaters—the males with artificially length-  ened tails—paid a heavy price. The added length apparently impaired their ability to fly. They could not hunt large insects, and their physical condition deteriorated: they did not molt well after the breeding season, and none of them returned from their winter migration the next spring, while many of the other males did return to breed in the same area.

Smith and Montgomerie[24] repeated Moller's experiments and found that males with either naturally long or artificially lengthened tails found mates earlier and started breeding earlier than other swallows. But when they tested for paternity among the nestlings by DNA fingerprinting, they found that only half of the nestlings in the nests of the inadvertent "cheaters"—the males with enhanced tails— were in fact the offspring of those males, compared with 95 percent in the nests of males whose tails were either naturally long or artificially shortened.[25] Smith and Montgomerie suggested that the glued-on tails were too much of a burden for males who were not fit to carry them, and that such males could not prevent their females from consorting with other males.

The experiments of Moller and of Smith and Montgomerie with barn swallows, and of Evans and Thomas[26] with the scarlet-tufted malachite sunbird, show the price paid by a bird who carries a longer tail than he can handle. Thus it would seem that longer tails, like the barn swallow's, the peacock's, and those of many other birds, are reliable indicators of the agility of strong, experienced birds and are attractive to females for precisely that reason.

Movements and Dances

The display flights and dances of male birds usually involve movements that are not common in their everyday lives. Turtledoves walk when they forage, but they hop around the female they are courting. Falcons flap their wings slowly in regular flight, but the male flutters his quickly in aerial courtship displays. The tropicbird even flies backward. In many species of songbirds the male takes on the burden of singing while in flight.

In some species whose males take no part in parenting, males spend a great deal of effort for many hours a day, many days in a row, on courtship dances. The females visit the dancing arenas, called *leks,* where they watch the performances and choose fathers for their offspring; very few of the dancers are chosen. A small number of outstanding males get most of the females, while young and low-quality males do not get a chance to copulate.

It can be difficult for human observers to identify the criteria by which the females select their mates—after all, humans often do not know enough to appreciate the difficulties involved in the particular dance. Gibson and his colleagues[27] showed that in sage grouse the best indicator of a successful male is a certain vocalization and a certain pause within it, which accompany a particular movement in the dance. Evidently, only the most superior males can achieve this particular combination within the strenuous choreography of the dance. As in human gymnastics competitions, in which most of the moves are part of a specific repertoire and are performed within a highly circumscribed framework, this very standardization is what enables competitors to best display subtle advantages and demonstrate the ability to execute difficult combinations of movements.

LEKS: CONGREGATING FOR DISPLAY

In most species males chase their rivals as far away as they possibly can, but in most lekking species males congregate, each in his own miniterritory within the lek, in order to display. Communal leks are found among insects, fish, amphibians, birds, and mammals, but only among species whose males do not participate in parenting. Among ruffs, many species of grouse, and several species of bird of paradise, males court and display in leks; among ruffs and sage grouse, hundreds of birds may congregate in one lek.

Lekking males do not have equal success in reproduction. Most males ultimately do not copulate at all; two or three males—who are usually in the center of the lek— may do over 90 percent of the copulating. Why, then, do the other males come together where they have little to gain? Probably because elsewhere their chances are smaller still. Petrie and her colleagues note that on days when many females visit the top male, other males may attract some females too.[28] Hoglund and his colleagues found that the more males in a lek of ruffs, the more females there are *in proportion to* males.[29] Females, after all, find it easier to choose a father for their offspring when they can compare males side by side. More than one female can afford to choose the same male; they do not mind sharing him with others because they will not have his help in raising their young anyway. Under such circumstances, it is in the interest of females to be able to compare as many males as possible, as easily as possible. Gibson and Hoglund found that young female grouse tend to prefer males selected by older, experienced females: they watch the experienced ones, then choose the same males.[30] If for these reasons females ignore small congregations of males and insist on the greater choice afforded by larger ones, males are forced to compromise with other males and congregate.

Manakins are small birds of the American tropics that display in leks.[31] Among them there are males who dance in a miniterritory within the lek in groups of two to five. Only the top male of each group, who is older than the others, copulates with the females who visit the courting arena. The others cooperate with him in the dance without immediate reward; but some of them can benefit in the long run. After his death, it seems that his top helper, who is usually at least six years old, inherits the top male's miniterritory.[32]

Female manakins seem to prefer the group perform- ers. No wonder: the top male in a group arena demon- strates both his dancing ability and the deference that other males pay him, which makes him all the more attractive to females. The other members of the group gain as well: they can both practice their dancing and increase their chances to inherit a good miniterritory.

POLYMORPHIC SPECIES AND MALES THAT MIMIC FEMALES

In leks of ruffs, one finds both dark-collared and light-collared individuals.[33] Dark- ruffed males fight for and acquire miniterritories within the lek, while males with

light ruffs move frequently from one territory to another, sharing it with the dark-collared male who presides in each.

The light-collared ruffs display subservience toward the dominant dark-collared ones by crouching with their beaks touching the ground when they approach, but even so, females copulate with the light-colored ones willingly. In fact, sometimes females seek out the light-collared males, even though nearby in the very same arena a dark-collared male is waiting.[34] Why do the dark-collared ruffs allow the light-collared ones to stay in their arenas?

Because females fancy light- as well as dark-collared males, it may well be that arenas with light-collared ruffs present attract females better than arenas without any; the dark-collared males therefore welcome the light-collared ones and proceed to "share the wealth." But why do females seek to copulate with the subservient light-collared ruffs?

Females who copulate with both varieties (morphs) of ruff are likely doing right by their offspring. The differences between the two morphs seem to be genetic. Since both morphs have evolved, it would seem that each has its own advantages. If these advantages could be combined in the same individual male, one would assume that natural selection would have merged the two morphs long ago. We are not familiar enough with the life strategy of ruffs outside breeding season to tell what the advantages of each morph are; but similar cases among fishes and crustaceans suggest that each morph is best fitted for a specific ecological niche. Both dark-collared and light-collared ruffs thus have a good chance in life, and it is to the female's advantage to have some offspring of each morph; she hedges her bets by having some offspring that are best suited for one life strategy, and others best suited for the other.

Adult male bluegill sunfish may be either small or large. The large ones defend breeding territories and court females; the small males join them during courting, and both fertilize the eggs laid in the large male's territory.[35] The color and movements of the small male are similar to those of the female. The large territorial males do not chase the small males away as they do other large males.

The common explanation for this behavior is that the small males "cheat" both the large territorial males and the females by pretending to be females themselves, getting access to a large male's territory and fertilizing the eggs females lay unbeknownst to either the females or the big male. But human observers do not find it difficult to distinguish the small "mimetic" males from females; why should it be difficult for members of the sunfish's own species, who after all have a lot more to lose by their "mistake"?[36]

We think that in fact there is no cheating involved, that rather this is a reproduction arrangement that satisfies all parties. The two male forms face different constraints of predation, feeding, and so on. Both survive and breed successfully. To have offspring of both morphs, which raises the odds that her offspring will survive and breed, the female's eggs must be fertilized by both morphs; therefore she may well prefer territories in which both morphs are present. Thus it is more likely that the large dominant male will attract a female if he accepts a small male as a partner. Since both females and small males display submissive behavior, each for its own reasons, it is not surprising that they use similar signals; after all, they are sending the same message. But that does not mean that the large territorial male or the female mistake the small male for another female. True, this is speculation; but the suggestion that the small male is cheating is also speculative, and a far less likely explanation in our opinion.

Similar "cheating" systems have been described in fish, reptiles, birds, and insects and are probably present in other forms of life as well. We believe that a close look at the facts in each case would reveal that each party to the system is actually straightforwardly serving its own best interests under given conditions, without deception. As we shall see in chapter 16, there are even social systems in which two species, host and parasite, cooper-

ate in order to reproduce. In the case of some of these interactions, too, one party has been described as cheating the other; and they too are far more interesting if one looks at them as forced collaboration between two parties with conflicting interests, each of whom tries to get the best it can. The result is a compromise between the opposing interests of the parties rather than a temporary triumph by one or the other through cheating.

FISHER'S MODEL OF THE RUNAWAY PROCESS COMPARED WITH THE HANDICAP PRINCIPLE

According to our theory, "waste" in sexual display is evidence of the advertiser's quality, but very different explanations have been offered ever since the days of Charles Darwin himself. The issue first came up in Darwin's *On the Origin of Species.* Darwin suggested in 1859[37] that the features and qualities of each species are formed by the process of natural selection, in which the more efficient survived and reproduced while others did not. This theory, though general and all-encompassing, did not explain the waste involved in the showing off that precedes sexual reproduction. Darwin could not see how an investment in showing off increases an individual's efficiency. He suggested, then, that there are two kinds of selection: natural selection, which makes an animal best suited to its environment, and sexual selection, which assists an animal in competing within its gender for the chance to reproduce.

Darwin[38] thus defined sexual selection as competition with members of the same sex. With this definition Darwin lumped together straightforward, efficient fighting between rival suitors with features that enable an individual to deter rivals of the same species and gender by means of threats, and with features that attract potential mates. Darwin did not see a problem in the evolution of bizarre signals that function in sexual advertisement—he simply turned his observations into an explanation. The simple fact that bizarre signals attract mates and deter rivals justified for him the investment animals make in these signals. He did not ask why waste attracts mates and deters rivals. Rather, he treated these effects as a given.[39]

In the early twentieth century, Fisher recognized the problem presented by animals' preference for wasteful signals. He stated, rightly, that female preference is an adaptation like any other produced by natural selection; he then asked why females prefer wasteful males.[40] The model Fisher proposed to answer this question used to be the only one available, and many still believe in it. The model can be found in almost any book on evolution.[41]

Fisher believed that the male who shows off is no better than the male who does not show off. According to that premise, the showing off itself is a drawback, and thus the show-off males are less well adapted than their fellows.[42] According to Fisher, the only advantage the ostentatious males have is the fact that females consider them attractive. Since such males pass on the show-off character to their offspring, those offspring will show off and will be attractive to females too. According to this script, males gain by investing in showing off because by showing off they attract more females. Females lose by having offspring who waste resources on showing off, but they have no other choice: only wasteful offspring will be attractive to other females, who in turn will have inherited from their mothers the tendency to be attracted to wastefulness.

Fisher's model can be seen as a catch-22, in which each individual male in the population wastes resources on showing off solely because this is the accepted method of courtship in his species. Several mathematical models support the idea that once some of the population consider a particular random feature to be attractive, then a "runaway" process may develop by which the feature spreads quickly throughout the population. In such a population, a female who went against the trend and selected a male who is more efficient and shows off less might have more efficient male offspring—but these offspring would find themselves without mates, since other females would not choose them.

Fisher assumed that the process starts when some females, who select a male by some feature that truly correlates to his general quality, get more and better offspring than females who mate indiscriminately. But in Fisher's view, once the daughters of these females inherit the tendency to select males by this feature, then males who exaggerate the feature—irrespective of their real quality—will get more mates. From that point on, according to Fisher, females choose males not by their quality in general but rather by the exaggerated feature, which in his view no longer correlates to the male's quality. In fact, Fisher said that the exaggerated feature decreases the real quality of the males, but that the process is driven forward by the preference of the females trapped in the catch-22. In other words, females now prefer males with the exaggerated feature because other females prefer them.[43]

But there is a major problem with Fisher's model. The same wasteful characteristics that attract mates also deter rivals of the same sex, and we have to ask why. If only females reacted to a feature that did not correlate to real quality, one might conceivably explain the value of having sons who bear the same feature. But we find that not only do these exaggerated features attract females, they deter rival males too.

If the signal indeed has no connection to the real quality of the male as a rival, then a male who is not deterred by the feature will succeed better than those who are. He will produce offspring who are likewise unfazed; eventually the arbitrary feature will die out as a threatening signal in the entire population. Yet this does not happen in nature; in the real world, rivals in many cases remain intimidated by the same supposedly arbitrary features that attract females, or by features similar to them. In fact, Fisher himself noticed this weakness in his model. But as Fisher had no way of explaining how display could be correlated to prowess, he suggested that with time, rival males will stop reacting to features that amount to mere "war-paint."[44]

By Fisher's model there need not be any correlation between the male's actual prowess and the female's choice, and indeed, Fisher assumed that the wastefulness of the signal decreases the male's true prowess. Yet many findings suggest that in fact the most extravagant males are also the fittest.[45]

The real question is not whether Fisher's model is internally logical or whether it can be expressed in mathematical terms, but rather how one can best explain ostentatious waste in nature, including showing off toward rivals. Any explanation must show why such high-cost signals as the singing of male songbirds, the large, heavy, branched antlers of male deer, and the colorful plumage and dancing of various male birds often deter rivals at least as effectively as as they attract females. In fact, Andersson remarks that it is often difficult to tell whether a given feature is used more for the former purpose or the latter.[46] Fisher's model explains neither the evolution of features used to deter rivals, nor the logical connection between signals and their messages.

Unlike Fisher's model, the theory we offer—the Handicap Principle—does explain the relationship between the specific way in which an animal shows off and the individual's quality both in courtship, where it attracts potential mates, and in competition with rivals, whom it deters. According to our model, the cost—the "waste"—is the very element that makes the showing off reliable. The female in this model is not a silly creature attracted to extravagant males just because "every female is"; by selecting as a mate a male who can afford lavish displays, she is choosing a good father for her offspring.

UTILITARIAN SELECTION AND SIGNAL SELECTION

We believe that natural selection encompasses two different, and often opposing, processes. One kind of selection favors straightforward efficiency, and it works in all areas except signaling. This selection makes features—other than signals—more effective and less costly; we suggest calling it "utilitarian selection." The other kind of selection, by which signals evolve, results in costly features and traits that look like "waste." It is precisely this costliness, the signaler's investment in the signals, that makes signals reliable. We suggest calling this process "signal selection."

Darwin included in sexual selection—the competition for mates—both signals on the one hand and features that actually enable an animal to fight more efficiently with rivals of the same species and gender on the other hand. Our definition, by contrast, makes a clear distinction between features that can be explained by straightforward utilitarian selection and those that cannot—signals. As we see it, most of what Darwin defined as "sexual selection" is better understood to be "signal selection."[47] Signal selection differs from sexual selection in that it includes *all* signals—not just those that affect potential mates and sexual rivals, but also signals sent to all other rivals, partners, enemies, or anybody else. At the same time, signal selection excludes features that improve actual fighting ability, which are selected for straightforward efficiency.

The need for reliability explains the multitude of signals in the natural world, and the theory of signal selection thus offers new ways of looking at every species on earth, from microscopic organisms to humankind itself. How signals evolve and what their evolution implies carry us through the chapters that follow.

PART II

METHODS OF COMMUNICATION

CHAPTER 4

THE FALLACY OF SPECIES-SPECIFIC SIGNALS

DID DECORATION EVOLVE TO IDENTIFY SPECIES, GENDER, AND AGE?

We have seen how signals used by enemies or rivals evolve: the engine driving this evolution is the common interest the antagonists have in reliable communication. But are there signals that evolve without the element of competition or enmity? What about those structures and markings that allow one to tell a given species from another, or to distinguish the male of a species from the female, and the juvenile from the adult? Scientists call these *species-specific* or *set-specific signals,* and they assume that common interest—an animal's need to identify other members of its own species—rather than competition among individuals explains the evolution of these traits. But we believe that specific signals evolve just as all other signals do—through the competition among individuals to demonstrate their quality.

To most people it seems obvious that species-specific signals exist so that we and other animals can tell one species from another. After all, birders in a new environment can identify the species, and often the gender and age, of unfamiliar birds by comparing the birds' decorative markings with pictures and photographs. We recognize the bulbul by its black head and yellow undertail coverts, the mallard by its green head and the blue on its wing, the blackbird by its glossy black body and orange beak; the markings of the male mallard and blackbird distinguish them

43

as well from the drab females of their own species. In breeding season, it is easy to tell the species of male ducks apart by their colorful plumage. Experienced birders can also distinguish them in other seasons, by subtle differences in the patterns of color on their wings. It can be very difficult to distinguish between closely related species: there are about ten species of stonechat and wheatear in Israel, and the small differences between some of the females are not easy to spot. Yet experts can tell even these apart by subtle, species-specific variations in their markings.

Lorenz[1] was impressed with the multitude of colorful fish in the coral reef. He suggested that their conspicuous markings help the fish recognize members of their own species so that they can avoid fighting with members of other species who are not competing with them for resources and mates and avoid courting and mating with members of other species. Wallace, in his argument with Darwin over sexual selection, proposed that the main function of male showing off is species recognition;[2] Mayr also suggests that most features unique to males developed for that reason.[3] This explanation is accepted by most researchers. After all, zoologists are used to recognizing species, gender, and age by animals' markings; they find it easy to assume that the markings evolved in order to aid animals in making the same distinctions.

Do animals really use markings to identify members of their own species? Experiments say that they do. Some species of gull have black heads and white rings around their eyes; others have black heads and no rings. When Smith[4] painted white rings around the eyes of the latter, other gulls treated them as though they were members of the species with the rings. When he painted black over the white rings of the former, they were treated by other gulls as members of the ringless species. Katzir[5] found that fish in a coral reef reacted to models painted to look like their own species and disregarded models painted with the markings of other species.

But did these markings evolve *in order to* enable animals to identify members of their species? Not necessarily. We identify a kangaroo by its special shape and gait. It is very likely that kangaroos themselves do the same. Yet no one suggests that the shape and gait of kangaroos evolved in order to help kangaroos recognize each other as kangaroos. The fact that features are *used* by animals to identify the species, age, or gender of other animals does not prove that the features *evolved for that purpose*.

In most species of babblers, the group-living desert bird we study, males and females look alike. Luckily for us, in the particular species we observe at Hatzeva there is a minute difference between the two: females have dark brown irises, while males have lighter, yellowish-tan irises. These birds treat strange babblers according to their gender: males attack males and court females, females attack strange females. One could assume that the difference in eye color developed so that babblers could tell the gender of other babblers. But with time we became skeptical: these birds showed us that they can tell the gender of a strange babbler long before they see the color of its eyes— perhaps by its voice (females' voices are higher), or by characteristics of its flight. If they can tell a stranger's gender from a distance, what is the point of one small gender-specific marker that can only be seen at close quarters? Would anybody suggest that women use eyeliner in order to let men know that they are women? Men can figure that much out long before they see a woman's face, let alone her eyes. We were thus led to the conclusion that the different colors of the male and female babbler's eyes evolved not to indicate gender but for some other reason.

And if the evolutionary value of species-specific markings is to prevent fights or to prevent courting among members of similar species who live in the same geographic area, then what is the usefulness of these markings in animals who live at the top of a mountain or on a remote island, far from similar species? Such species retain their specific markings even after long periods of isolation.[6] Mutations of color and markings are not unusual among animals in captivity; there is every reason to assume that such mutations occur in the wild as well. We rarely see them, because they are selected out. Distinctive markings thus have a selective value even in isolated areas; otherwise natural selection would not have preserved the species-specific markings so faithfully.

Some of the most prominent and colorful decoration is found in male birds of species that are not monogamous, yet it is precisely these males that are especially prone to mate with females of other species. Selander, who remarks on this fact, explains it by saying that the polygamous males do not lose much by fathering a few infertile offspring among their many fertile ones. But for a female, even in polygamous species, infertile young represent a substantial loss. Raising infertile offspring demands as much effort on the part of the mother as raising fertile young does. Thus a significant part of her reproductive potential ultimately goes to offspring who cannot themselves reproduce.[7] If species-specific decoration and markings evolved for the purposes of species identification, why do mistakes happen more often precisely in polygamous species, whose males have the most distinctive markings?

Species-specific markings raise a much more basic question, however: if the only purpose of such markings is to enable animals to identify their own and other species (or gender or age within their species), it is difficult to imagine the process by which they evolved. Markings can help establish group membership only if they are common to all members of the group, or at least to many. Yet the first individual who bore those markings was a minority of one. In order for the markings to become "species-specific," their bearer or bearers must have reproduced better than individuals who did not carry the same markings, until with time the trait spread throughout the population. The early bearers of the markings had to have some advantage over individuals who did not bear them. That advantage could not be that it was easier to identify them as members of the species, for at that point most members of the species did *not* carry the markings.

THE EVOLUTION OF COMMON MARKINGS THROUGH COMPETITION BETWEEN INDIVIDUALS

We suggest a very different explanation: these markings evolve through the *competition* that members of the species engage in to determine their relative quality.[8] The markings that we see as *uniform* are the very ones that show most clearly the fine *differences* among individuals regarding the attributes most important to them. The closer the competition—the more evenly matched individuals are with regard to some desirable attribute—the more helpful uniform markings can be in exposing these fine distinctions. Anybody who has ever had to judge an athletic, musical, or beauty contest knows how crucial it is for the athletes to compete under highly regulated and calibrated conditions, for the musicians to play under similar conditions, or for the beauty contestants to appear in similar clothing, precisely in order to tell the fine differences between them and select the best of the best.[9]

Tourists in the old city of Jerusalem, especially those equipped with a good guidebook, can identify members of many different groups by their clothing: nonobservant Jews; ultraorthodox Jews, and even members of particular Hassidic communities; Arab villagers from different regions, who dress differently from each other and from Arab city dwellers; Bedouins, whose clothing is different yet again. The tourist might think that people wear specific clothing to avoid the risk that, say, an ultraorthodox Jewish girl might unwittingly marry a Bedouin or, God forbid, a nonobservant Jew. But any local can easily tell which group a person belongs to, regardless of that person's clothing: by his or her gestures, speech, movements. If a Bedouin were to put on ultraorthodox Jewish garb, an ultraorthodox Jewish girl would immediately spot him as a pretender.

Why, then, do people of a given group make the effort to dress "properly"? Why does an ultraorthodox Jew dress just like all the other members of his synagogue? Or, for that matter, why do corporate executives, or members of a biker gang, dress just like other corporate executives or members of the gang? The clothing within each group is not really identical, of course. Small details—the quality of materials or workmanship, the fit, the care taken in dressing, the way a person carries his or her clothing—vary from person to person, and the differences tell a great deal about each wearer's means, abilities, and personality, as well as accentuating posture and behavior. It is precisely the similarity of the clothing that makes it easier for group members to assess the differences among them. If each person within the group wore wildly different clothing it would be difficult to compare them with one another in a meaningful way.

How exactly can uniform decoration show the difference between individuals? Let's start with a simple example: let's say we are going to the marketplace to find a handmade round plate. The plates are not identical. Some were made by skilled craftsmen and are perfectly round; others were less skillfully made and are slightly elliptical, or their edges are a bit malformed. A small circle in the middle of a plate would make it easier for us to tell the most 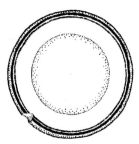 perfectly round plates from the less perfect ones. A line around the edge of a plate would make it easy to spot imperfect edges. If we wish to select perfectly round plates with neat, perfect edges, it would make sense for us to select from among plates decorated with a circle in the middle and a line around the edge, even if we have to pay a somewhat higher price for them. Since other shoppers are also looking for well-crafted round plates, it would make sense in turn for the skilled artisan to paint a circle in the middle of each plate and a line around its edge.

This decoration will clearly benefit the makers of perfect plates, but why should the makers of less than perfect plates decorate them with a pattern that will bring out their imperfections? They should do so because they are competing with yet other craftspeople, whose plates are even worse. Buyers who cannot find or cannot afford the best plates would thus be reliably drawn to the better ones among the ranks of the imperfect. Even the makers of the absolute worst of the plates had better invest in such patterns, since buyers will reject out of hand plates that don't bear the pattern, and since there might appear an incompetent artisan making even worse plates than theirs. The process is driven by the preference of the buyers.

This issue of choice is crucial. An essential condition for the development of any signal is that the receiver of the signal must have at least one other alternative— otherwise there is no chance of affecting the receiver's actions and no point in signaling to it. A gazelle can tell a wolf reliably that it is an excellent runner and that the wolf has little chance of catching it, but if the wolf doesn't have the alternative of finding other prey, it has to try to catch that gazelle, even though the gazelle has just proven that the chase will be a hard one.

If, as we believe, animals' markings evolved in order to bring out small differences between competing individuals within a species, then the markings cannot be arbitrary. Not every decorative pattern would enable us to judge a plate's roundness equally well; many patterns would actually disguise imperfections and make it more difficult for us to find the better-made plates. Likewise, any markings could serve to identify species, gender, or age. But if species-specific markings show off certain important traits in members of the group who compete to prove their quality, the markings that evolve have to be those that best bring out differences regarding these traits. And indeed, once one starts looking at animals with this in mind, examples abound.

MARKINGS AND THE TRAITS THEY ADVERTISE

Lines and Stripes

Barlow[10] noted that fish with unusually shaped heads have decorative markings that conform to the shape of their heads. We likewise find that long fish generally have lengthwise stripes, and high-bodied fish generally have vertical stripes. Barlow assumed that the stripes camouflage the fish. Our own experience is that striped fish are easier to spot. Lengthwise stripes bring out the length of the fish, or to be more exact, make it easier to spot slight differences in length among fish of the same species. Vertical stripes bring out the height of the body, and a center stripe brings out symmetry of form.[11] The pattern clearly draws the eye of the beholder to whatever is the object of the decoration.

Species of coral fish that are short, high-bodied, and thin are decorated with vertical stripes. Moreover, the vertical stripes emphasize specific proportions: sometimes the line starts at the front corner of the dorsal fin and goes straight down, sometimes it runs from the back of the dorsal fin to the front of the anal fin, and sometimes it runs across the base of the tail, or across the head, through the eye. Since the lines run through such important body parts, they cannot lie: a line that was longer because it had a different slant would be obvious, since it would not then go through the same features.

The paragon of stripes is, of course, the zebra. The zebra's stripes can be seen clearly only by nearby observers: from farther away they merge into a discreet gray.[12] If the stripes were meant purely for camouflage, they could have been random, like a leopard's spots. But a zebra's stripes emphasize specific body parts: lips and hooves are black, crosswise lines show off the thickness of legs and neck, and other lines show off the shape of the rump.

Patches and Frames that Bring Out Body Parts

Many species of hoofed animals have markings that outline their rumps: patches of color, as in most antelopes, or stripes like the zebra's. We realized the value of such markings when we traveled with a student researcher at the Shushlui park in South Africa. At the time he was studying waterbucks, large antelopes with white rumps. Many of the waterbucks were infested with ticks and were thin and weak. Their white rumps made it easy to tell from behind which antelopes were in good shape and which were thinner than they should be. On thin antelopes, the white patch had the shape of a pointed ellipse because of the atrophied hind leg muscles; on healthy animals, the patch looked nice and round. The hind leg muscles are of immense importance to the antelope—they are its "engine," its driving force. A predator looking for easy prey, a rival evaluating his chances in a contest, a female looking for the best father for her offspring, all can benefit from evaluating the muscles of the rump.

The wings of many butterflies are outlined with a narrow frame of color. If the wing were uniform in color, it would be difficult to spot small defects in the edge. The colored outline, on the other hand, is visibly broken by imperfections in the wing's shape. Such breaks can result from developmental  defects, encounters with predators who take bites out of the wing, or collisions with hard objects. They are more likely to occur in older individuals. The outline can enable females to spot broken edges and so avoid mating with males who have birth defects, are clumsy, or have reached the age where they have spent most of their sperm. Similar frames and colored edges on feathers enable birds to tell the condition of those feathers at a glance.

Beaks, nails, hooves, fins, spines, tails, and horns are often a different color than the body. The difference brings out the shape, size, and movement of these parts, just as among humans nail polish shows off the exact shape and movements of nails and fingers.

IMPLICATIONS OF THE HANDICAP PRINCIPLE: THE USE OF MARKINGS TO FOCUS ON FEATURES

Once we realized that there is a relationship between an animal's decorative markings and the information those markings convey, we started looking at animals differently. Till then, we had been looking at animals and trying to identify their species. From then on, we have been trying to see what an animal's markings can tell us about its features and adaptations. Which body parts are emphasized by an animal's decoration, and what is their importance to that animal? Again, we found that this new approach revealed to us many features that we had never noticed before.

The yellow dot near the base of a surgeonfish's tail emphasizes the spike that is located there. The stripe above the eyes of a trunkfish accentuates the special shape of its head. The colored dots at the tips of mallards' and pelicans' bills

decorate hooks at the end of their beaks: we never even noticed the hooks until we started looking at markings as signals. The red patch near the tip of the beak on the lower mandible of the herring gull shows off the exact dimensions of a thickening there, which is typical in larger gull species. Before we started looking at markings as meaningful in this way, we never noticed the thickening, or remarked that there is no such thickening in the beaks of smaller gulls.

We are not the first to observe the dot on gulls' beaks. Tinbergen[13] noted that herring gull chicks peck at the colored dots on their parents' beaks, and that the pecks move the parent to regurgitate the food it has brought its young. Many ethologists saw this as the reason the colored dot evolved. But Tinbergen's argument does not explain the exact location of the dot; and the young of smaller gull species, which don't have a colored dot on their beaks, also peck at the tips of their parents' beaks. We think rather that the colored dot evolved to show off the thickening at the tip of the beaks, and to emphasize small differences between individuals regarding the size of this thickening; its use by chicks is secondary.

The flight feathers on a stork's wing and the large feathers that cover them (coverts) are black; the small coverts, which make up the rest of the wing's surface, are white. The wing's aerodynamic properties depend, among other factors, on the thickness of the various parts of the wing. If the wing were colored uniformly, it would be difficult to see the dimensions of its parts in relation to one another. The two colors show off not only the general shape of the wing but also the relative size of its main parts and changes in their shape during flight, and thus the quality of flight.

Bustards, stone curlews, and many butterflies are very well-camouflaged on the ground; in flight they display the conspicuous patterns on their wings. The patterns help onlookers assess the speed and direction of the individual's flight, the maneuvers performed, and so on.

Many birds' necks are a different color than their bodies. The position and movement of the neck tell a great deal about the intentions of the bird. Is it calm? Is it on the verge of flight? Is it tense or frightened? The contrasting color of the bird's neck shows off both its intentions and the length of its neck.[14]

Fish's fins often have eyelike patterns on them. The pattern brings out movements of the fin: when the fin is held down, the round "eye" changes into an ellipse or even into a line. A fish ready for quick movement folds its fins down; when threatening, a fish spreads its fins, and the "eye" shows up as a perfect circle.

Color may also be related to health. The rooster's comb, the turkey's featherless head, humans' lips and cheeks, the small patch of bare skin on the forehead of the chick of the great crested grebe, all vary in color depending on the amount of blood they receive: brighter in healthy individuals, these areas turn pale and even bluish when the blood vessels narrow in reaction to cold, ill health, or other stresses.

Hamilton and Zuk[15] suggested that bright colors tell female birds that a male is healthy and is not suffering from parasites. According to them, it pays off for females to choose a brightly colored male, since the bright colors show off its genetic tendency to withstand parasites—a tendency it would pass on to its off-spring. Hamilton and Zuk's article led to many research projects, some of them showing a reverse correlation between bright colors in birds and the number of parasites the birds carry. In fact, the point can be made much more sweepingly: stress created by parasites is no different from stress caused by cold or hunger. The brightly colored plumage of a male in good shape develops properly; the plumage of a sickly one cannot. Indeed, Hamilton and Zuk refer to the effect they describe as a special case of the Handicap Principle.

Markings can also indicate how healthy a bird was when it grew its feathers. Birds molt at least once a year; individuals in ill health do not molt, or molt only partially. Patterns typical of new feathers—for instance, narrow rims of color that wear off in time— give proof that the bird's plumage is new. The plumage of a bird who was in ill health during the molt is often partly made up of old feathers; decorative patterns on new feathers make this readily apparent. Many birds have feathers with a line down the middle; missing feathers are easy to detect, because they disturb the pattern of the combined lines.[16]

SYMMETRY

The correlation of symmetry to quality, both in animals and in humans, has been studied recently by several researchers; the term used in current literature is "fluctuating asymmetry." Moller[17] found that the male swallows preferred by females had both longer and more symmetrical tails (that is, the left and right parts were evenly matched) than males who took longer to find mates. Research confirms that humans consider symmetrical faces to be more beautiful than somewhat asymmetrical faces.[18] Thornhill[19] found that male scorpion flies who won fights with rivals, and who brought more presents to their mates, had more symmetrical wings than the males they defeated. It is not clear whether female scorpion flies actually chose males for their symmetry; even when the females selected males by smell alone, without seeing them, the ones they chose were more symmetrical than the ones they passed up. What, then, is the connection between symmetry and quality?

Stress or genetic flaws often cause asymmetrical development of body parts.[20]

If the problem were simply lack of nutrients or of energy, one would expect that deficiency to affect both sides of the body evenly and cause, say, a shorter symmetrical tail rather than a longer, asymmetrical one. We believe that symmetrical growth requires reliable communication within the body.[21] For synchronized, symmetrical growth, the center that regulates growth must send the same messages to both sides of the body, get feedback on the results, and pace further development according to the data received.

As we shall see in chapter 9, signaling between cells in the multicellular body requires reliability no less than communication between organisms does. It seems that only a cell in good shape can afford to manufacture a quality signal. In other words, reliable communication within the body demands an investment on the part of the organism, and it may be that under conditions of stress the body cannot spend enough on such communication to ensure symmetry. Thus a symmetrical shape indicates better overall health during development than a less symmetrical shape.

An awareness and understanding of "esthetic" features such as symmetry, good color, and patterns that suit what they decorate thus bring real and concrete benefits to those who possess them. Indeed, in the last chapter of this book we will suggest that this connection between "beauty" and quality is the basis for the evolution of the esthetic sense in humans.

The right decoration can reveal the degree of symmetry.[22] A line down the middle of the body, or down the middle of a body structure, is an obvious example. A circle in the middle of an area that ideally is symmetrical is another. In fact, any symmetrical decoration on body parts helps observers assess symmetry of shape and structure; and since the decoration cannot be more symmetrical than the body parts themselves, the information conveyed is reliable.

"EYE" PATTERNS

The peacock's long tail feathers take several months to grow. The scores of tail feathers are arranged so that eyes form arches; the arches are very orderly, with each eye exactly halfway between two eyes in the arch above it. Missing eyes are very conspicuous—there is no need to count, and a peacock cannot hide the loss or imperfect development of even one tail feather. And as Petrie found,[23] the response of peahens is measurably different if even only five eyes are missing in a peacock's tail, out of a hundred fifty or more.

For a round eye pattern to emerge on a lengthening feather, the entire growth and development of that feather have to be in perfect synchronization. Any irregularity in development would ruin the perfection of the circle. A circular pattern—and even more, a circle within a circle—is thus ideal for displaying regularity in the process of growth. If the peacock's tail were decorated with lines, it would be difficult to be sure of deviations: a line of one length is not more perfect than a longer or shorter line. But an imperfectly developed circle exposes any distortion, in any direction.

This value of eyelike patterns can explain their wide-spread occurrence, as on the wings of butterflies, mantids, and other insects. The standard explanation—that the eyes frighten off predators, who mistake them for the eyes of large animals—is an insult to the predators' perception; nor does it explain patterns composed of many eyes, or of series of four or five concentric circles. We doubt that any predators are frightened off by the eyes on a peacock's tail.[24]

THE EVOLUTION OF MARKINGS; POLYMORPHISM AND CONVERGENCE

If indeed markings are not arbitrary, and certain markings show off specific qual-ities better than others, then that explains how a set of markings can spread throughout a population. First, members of a population start focusing on a par-ticular feature—the size of a beak, say, or certain movements—because they find that the feature in question conveys essential information either about the overall quality of other individuals, or about whether they possess certain qualities. Once that happens, then individuals decorated with markings that bring out these fea-tures, other things being equal, are more likely to be selected as mates, or avoided as rivals. They have a selective advantage over others; they will have more descen-dants than individuals who do not bear those markings. Thus, the markings will spread in the population and eventually become "species-specific." When mem-bers of a species must compete to show off their superiority in qualities important to the life of that species, individuals who can demonstrate such superiority have an advantage. The interest in the size or shape of a beak, rump, or tail leads to decorative markings that bring out reliably the size and shape of that beak, rump, or tail.

The markings of closely related species and even of subspecies are often dis-similar. A new species or subspecies usually evolves when a population of a certain species adapts to an environmental niche that is somewhat different from the orig-inal niche of that species. This can require physiological, morphological, or be-havioral adaptations.

The new niche can change the relative importance of certain features—like, for example, beak size.[25] When this happens, it makes sense that the features most important to members of the "new" species are different from those most impor-tant to the original species. The decoration best suited to bring out these features will therefore also be different. Individuals with decorative markings that best emphasize these new characteristics will have an advantage over individuals with the old decorative markings, and the new markings will spread in the population. First comes the evolution of new adaptations; next, individuals start focusing on these adaptations; and finally, because of that attention, new decoration emerges— the new species-specific markings.

But the story is not always this simple. Sometimes members of the same species have different markings. One such polymorphic species is the ruff mentioned in the previous chapter. Some male ruffs have dark ruffs, others have light ruffs. The courtship behavior of each type is also different. The dark-ruffed males tend to hold a territory within the lek and dance in it, while the light-ruffed males tend to fly from one territory to another, dancing with several of the dark-ruffed males in turn. As we discuss in detail in chapter 8, black and other dark colors better emphasize size and shape, while white is better for showing off movement. The darker color of the territory-holders better advertises their stance, while the fact that the light-colored ruffs are so often in motion explains why they advertise themselves with lighter colors. We don't think that the difference in colors evolved so that what we now call the dark and the light forms could be distinguished; rather, we believe it evolved to enable each light-colored ruff to try to show himself as the best among the light-colored ruffs, and to allow each dark-colored ruff to compete as successfully as possible against other dark-colored ruffs.

The relationship between decoration and its message explains why it so often happens that unrelated species who occupy a similar ecological niche look alike. The decoration of the desert agamidae of the Middle East is very similar to that of desert iguanas in the Americas, as is the decoration of rattlesnakes in America and that of vipers in the Old World. Lack remarked that birds that live in similar conditions tend to have similar decoration.[26] In the Argentinean plains, we encountered what we thought was a crested lark—only to learn that it was in fact a larklike brushrunner (*Coryphistera alaudina*). Species that live in similar environments tend to evolve similar traits to best deal with these environments; they then advertise these traits with the markings that best show them off— which tend to be similar patterns and colors.

FACIAL MARKINGS AND THE DIRECTION OF GAZE

Most animals have facial markings. Often these markings decorate the eyes and help onlookers detect eye movement and determine the direction of the animal's gaze. Birds often have eye rings, eye stripes, or colored lines running from beak to eye. Bird species with binocular vision, such as shrikes, stonechats, and great tits, often have eye stripes converging toward the beak. When both eye stripes look equal to the observer, the beak—and the bird's gaze—are aimed straight at him or her. Any change in the direction of the bird's glance will change the symmetry of the lines. The cream-colored courser, whose unusual field of vision extends backward, has decorative lines going from its eyes to the nape of its neck. Bulbuls, partridges, gulls, babblers, and the like have monocular vision, that is, each eye has a separate field of vision; these species often have eye rings. The bulbul has a black head and a black eye with a prominent white eye ring; the white ring enables a distant onlooker to tell the direction of the bulbul's gaze.

Lines, eye brows, and colored eye rings make it clear even from a distance what direction an animal is looking in and demonstrate his or her ability to stare—which, as we saw in chapter 2, is one indication of his or her level of confidence. This information is reliable: the same decorative pattern that emphasizes confident behavior brings out the hesitant movements of individuals who have not yet decided what to do and are still assessing their options. The direction of a courting or threatening individual's gaze indicates its interest in what it is looking at.

The eye decoration of youngsters, and often of females, is less prominent and revealing than that of adult males; for example, young bulbuls lack the white eye ring that characterizes the adult of the species. While youngsters do benefit by showing in a general way what they are interested in, they are better off if they avoid committing themselves—as we shall see in more detail in chapter 18.

STATUS BADGES OR HANDICAPS?

We have seen how the deer's large, heavy antlers and the peacock's heavy tail enable them to advertise their quality reliably. Not surprisingly, such tails and antlers affect how much recognition—prestige—others accord an individual. Although they are less conspicuous, decorative markings can also advertise social status. The great tit's black band, the white patch on a magpie's wing, the black patch under the beak of a Harris's sparrow, an eyelike spot on a fish's fin, and the bulbul's prominent white eye ring all serve to proclaim their owners' prestige (social status) and in fact are termed "badges of status."[27]

For example, great tits with a wider black band on their underparts defend their territories and breed more successfully than those with a narrower band;

great tits with narrower bands subordinate themselves to ones that bear wider bands.[28] Experiments in other animals confirm the value of such badges of status. But how can a purely decorative marking be a reliable indication of anything? Evolutionarily speaking, what is there to prevent a young or low-quality bird from developing a wider band and gaining an advantage over its more capable rivals?

Several researchers have run experiments in which they decorated younger individuals with badges of higher status. In some of these experiments, the status of such individuals rose accordingly, but this had costs. Rohwer and his colleagues[29] painted low-status Harris's sparrows with patches of color indicating higher status. They found that the "pretenders" not only did not gain but actually fared worse than young, unaltered control birds: they were attacked more often by high-status birds, which usually don't bother to attack youngsters. When researchers gave the painted birds injections of testosterone—which increased their aggressive behavior and reduced their hesitation—the birds did gain in status, but they may have lost more than they gained; their bodies seemed to suffer from the artificially higher levels of testosterone.[30] The researchers asserted that the increased aggression toward "impostors" keeps the markings honest and prevents individuals from putting on status badges they have no valid claim to. But if status is determined by an arbitrary tag, how do the birds recognize an "impostor" who bears "fake" badges that do not fit its "actual" status? And why should that specific marking have evolved as a status badge in the first place?

Status badges, like other markings, amplify differences in features that directly indicate an individual's quality. The status badge emphasizes the excellence of an individual of higher quality; the same badge brings out the inferiority of a lesser individual. Eye stripes emphasize head movement—and thus accentuate both the steady gaze of a confident individual and the hesitancy of an insecure individual. The confident animal gains, by drawing attention to its gaze, but the insecure one loses. The black band on a great tit's breast makes its breast look narrower than it is; the wider the band, the narrower the breast appears.[31] Only a high-quality great tit can afford the narrowing effect of a wide band.

Decoration that seems at first glance merely to tag its owners as being of higher or lower status is really a standard measuring tool that enables individuals to show off their quality—not as compared with those who lack the status badge, who are clearly subordinate, but rather with individuals who *bear the same marking*. Such patterns are not physical handicaps, yet they are handicaps all the same, since they impair the ability of lower-quality individuals to falsely present themselves in a way that would assist them in social conflicts. In other words, like all signals, these handicaps are more of a burden to lower-quality individuals than to higher-quality ones.

In sum, we believe that those markings that best display differences in quality or motivation among similar individuals evolve to become common to the entire species, or to an age and gender group within the species, for precisely that reason. Since different species have adapted differently, the decoration appropriate to those species would likewise be different. That others use these differences to identify the group to which an individual belongs is a side effect of the competition among individuals within the group.

ARE THERE SIGNALS WITHOUT HANDICAPS?

Hasson claims that some signals, which he calls "amplifiers"—mostly decorative markings—are not costly in themselves but simply show off the quality of the signaler without imposing any handicap on it. For example, lines along a fish's body or a feather's edge show clearly and reliably the length of the fish or the condition of the feather without any significant material cost to the fish or the bird.[32]

We do not believe that the cost of decorative markings can be separated from the dimensions of the structure they advertise, or from other messages they may carry. The line along a fish's body shows how long the fish is; that is its "message." Each individual fish's line bears a different quantitative message. The line lets a fish show off clearly how much longer it is than shorter fish—but once it has the line, it also can't avoid showing off clearly how much shorter it is than longer individuals. Thus the cost of the decorative marking is differential rather than uniform and is greater for inferior individuals than for superior ones. The material investment in the line may be the same, and may be insignificant, but the message and its cost are different for different individuals. The same markings that show off clearly their bearer's advantage over inferiors also impairs that animal's ability to "fudge" and pretend to be as good as individuals superior to it. The higher the quality of the individual, the more likely that those it is compared with will be inferior to it, and the more it benefits by bearing the markings; the lower the individual's quality, the more likely it is to be compared unfavorably with better-quality individuals, and the more costly the markings.

ARE THERE ANY SIGNALS THAT ARE CONVENTIONS?

Are there conventional signals that do not have any competitive aspects? Maynard Smith[33] assumed that specific signals, such as those proclaiming that an individual belongs to a certain species and is of a particular gender, are signals that convey noncompetitive information; comparing them to railroad timetables, he termed them "notices," and asserted that in the case of such signals there is no reason to cheat, and thus no reason to prove their reliability and no investment required of the signaler. Grafen, who developed a mathematical model that supports the Handicap Principle, also assumed that some noncompetitive signals exist; for example, relatives may communicate by means of agreed-upon, conventional signals, and it was his contention that in such a case, guaranteed reliability and the necessary investment would not be issues.[34] Many who today accept the principle that the cost of a signal guarantees its reliability do not yet accept our claim that *all* signals have a cost—they impose a handicap—and that this is what guarantees that they are reliable.

"Railroad timetable" signals or notices are indeed purely arbitrary and have no connection to their message or to cost: it is enough for the signal to be agreed upon, clear, and efficient. But we cannot find such signals in nature. The examples in this chapter all show a clear relationship between even small, unobtrusive signals and the characteristics they advertise; all these signals amplify the ability of the observer to spot superiority or defects in the animals that carry them. The more flawed the individual, the costlier the pattern. This uneven cost is characteristic of signals that arise through signal selection, all of which involve investment in reliability.

As we shall see when we discuss chemical communication, even in the case of communication within the multicellular body, where there is no conflict between communicating cells, reliability is needed—in this case to prevent errors—and can be achieved only by means of handicaps. As far as we know, there are only two biological systems having "agreed-upon," arbitrary signals that do not involve investment in reliability: human language—on which we will say more in chapters 6 and 18—and, according to the present consensus, the genetic code. But in fact, some believe that the genetic code is not arbitrary either.[35] Yet our assertion that there are no "ordinary," agreed-upon signals in the animal world is still contrary to the opinion of most researchers.

THE DEFINITION OF SIGNALS; INFLATION AS A TEST OF THE THEORY OF SIGNAL SELECTION

But what, after all, are signals? We define *signals* as traits whose value to the signaler is that they convey information to those who receive them. There is a difference between traits that evolved for other reasons, such as body size or a kangaroo's gait, that can and do convey information; and signals, which evolved solely to convey information. For example, one can judge the direction of another's gaze by watching its eyes—yet eye movement is not primarily a signal. But eye rings and small tufts of hair or feathers whose only function is to show an observer the direction of an individual's gaze more clearly, or from a greater distance, *are* signals.

The ability to observe and understand signals is an adaptation like any other that evolves through utilitarian selection. An individual who pays attention to unreliable signals will be less successful—will have fewer descendants—than one who insists on paying attention to reliable signals only. As we have seen, reliability demands investment by its very nature—the signaler's investment in the signal is the guarantee of its reliability.

It is not always easy to judge whether a specific trait is purely a signal or whether it has some other function, but it is still important to make the distinction. What, then, is the fundamental difference between signals and other traits? It is the relationship between the signal and its cost. Every trait, whether a signal or not, demands an investment of some sort, and every trait is constrained by other traits whose requirements conflict with it. In this signals are similar to all other traits. Changes in the environment may change the constraints on the evolution of traits and lessen or increase the cost involved. For example, the size of an animal may be limited by its need to stay small and quick to avoid predators, but if its predators are removed it can afford to increase in size and thus increase its ability to store energy, withstand cold, or defeat rivals of its own species. The species then becomes larger because the cost of being big has gone down significantly.

If, on the other hand, the cost of a *signal* is reduced to the extent that every individual can use it equally well, then the signal can no longer reveal differences in the quality or motivation of individuals. In such a case, the signal *loses its value.* Because the signal's cost has gone down significantly, the signal is no longer useful and will disappear.[36] The evolution of signals—signal selection—is thus fundamentally different from the evolution of all other adaptations.

In traits other than signals, the cost of the trait is an unavoidable side effect. In signals, cost is of the very essence; it is necessary to the existence of the signal. If there is no cost, nothing prevents cheaters from using a signal to their benefit and to the detriment of the receivers, and that signal will lose its value *as* a signal. This has happened not infrequently in human history. When money is easier to get, it loses value through inflation. Ornaments that were prized as tokens of wealth when rare became worthless when easy to obtain.

When lace was made by hand by expert workers, the amount of skilled labor needed to produce it made it very expensive; lace cost—and was worth—its weight in gold to the rich and powerful, who wore it to display their wealth. The development of machines that could cheaply manufacture lace indistinguishable from the handmade product[37] put an end to the use of lace as an indicator of wealth; today, in fact, lace is not much used at all.[38] By

contrast, commodities like bread or iron that are used to meet direct needs rather than to send signals do not lose their usefulness if their price goes down; rather, they are used all the more.[39] Although human cultural and economic development are different from biological evolutionary systems, the two have a very important common element: both emerge from the need competitors have to cooperate and communicate.

The significance of cost is unique to our theory of signal selection. No other theory of signal evolution—and certainly no theory that assumes the existence of agreed-upon, conventional signals—would predict that if its cost becomes low, a signal will lose value. It should be possible to test this hypothesis: even though experiments with evolutionary processes require many individual subjects and a long period of time, they are feasible with unicellular organisms or organisms that multiply rapidly.

Meanwhile, we can test our theory against observations of animals in the field. As this book was going to press, we learned of some findings that support our view of the impact of inflation on the effectiveness of a signal. In Australia, male satin bowerbirds especially favor blue objects; in their usual habitat such objects consist of blue feathers, which are rare, and blue flowers that need constant replacement. Borgia[40] found on average five blue feathers per bower. Near human habitats or picnic grounds, however, blue artifacts—mostly blue plastic—are fairly common, and satin bowerbirds can collect as many as a hundred per bower.

In most areas, male satin bowerbirds compete both by stealing blue objects from one another and by destroying each other's bowers. Hunter and Dwyer[41] recently found, however, that where blue objects are abundant, the males direct *less* effort into stealing such objects from one another, and more into destruction of their competitors' bowers, than they do in areas where such objects are rare. Blue objects, where abundant, are no longer significant marks of quality. The birds still collect them, but they seem to place no more value on them than humans do on cheap lace.

CHAPTER 5

MOVEMENTS AND RITUALIZATION

Movements can indicate many things: a scolded dog lowers its ears and tucks its tail between its legs, while a happy dog wags its tail; a threatening cat arches its back and stands on tiptoe, hair raised; a duck moves its bill up and down to signal its intention to fly; male ruffs court females by dancing in special dancing arenas, or leks. These are only a few examples of the endless array of movements used as signals. The Handicap Principle gives us new ways of seeing such signals and of understanding how they evolved and the messages they encode.

DIFFICULT MOVEMENTS

Male courtship movements are often extremely elaborate and difficult to perform. The superb bird of paradise hangs upside down on a branch, flapping its wings and spreading its ornamental feathers. Hanging upside down does not make the movements any clearer, but it does make them more difficult to perform, and thus stronger proof of the male's physical ability and his motivation to court females. Ruffs, manakins, and bustards perform elaborate dances very different from their day-to-day movements. Doves usually walk when on the ground, but when courting, male doves hop on both legs. The hop is probably more difficult and demonstrates the male's prowess.

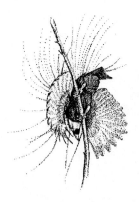

Male larks hover high in the sky to sing. At the song's end the lark does not fly down but rather folds his wings and plummets like a stone, opening the wings only at the end of the dive. Some individuals open their wings several times before touching the ground; others dare to make a breathtaking, uninterrupted dive, spreading their wings only at the last second. The later the wings are spread, the more impressive the dive. This is very similar to competitions among free-falling human sky divers: a wrong estimate can result in a disastrous crash. Differences in the dive may well reflect the relative expertise of specific male larks and convince females and other males alike of the performing bird's abilities as a mate or a rival.

Not all movement signals are meant to convey information as important as what an individual is worth as a mate or a rival, and the investment in other such signals can correspondingly be more modest. A dog signaling friendship wags its tail; this is a reliable indication of friendliness, since the wagging tail would interfere with attack or flight. A hungry nestling stretches its neck, beak gaping; a satiated nestling does not. The parent prefers not to exert itself to bring unneeded food; it demands that a nestling prove its need by making an effort too great to be worth making unless the nestling is indeed hungry.[1]

RITUAL FIGHTING

Much of the fighting that goes on between rivals of the same species is ritual fighting; the goal is not to injure or kill the opponent, but to convince it and others that it is the weaker and should withdraw from the contest. The initiator of the fight does not even try to surprise its opponent; rather, it signals clearly that a fight is imminent and lets its rival prepare itself; only after its opponent has assumed the appropriate stance does the aggressor strike—and then only a well-protected and "legitimate" target, such as antlers.

The horns of a male gazelle are strong, sharp, lethal weapons. A male gazelle fighting his rival could surprise it and run it through, but he does not. He comes at his opponent slowly and lets it get into position before locking horns with it and pushing and pounding it in a contest of strength. The end of such a grueling struggle usually comes when one of the opponents concedes.

It would be a mistake to call such a struggle a fight. It is more like a competitive sport in which contestants try to show off their superiority while following fixed rules. In fact, these are threats that take the form of a physical contest; the winner does not pursue the loser and kill it; it is satisfied when its rival gives up. In the rare case when one of the contestants dies, it is because it failed to withdraw in time.[2] Male gazelles hand-raised by humans as pets treat people as members of their own species, and when grown up they may invite their keepers to a rivals' contest. The keeper, who does not realize that the gazelle's initial butt is an invitation to ritual combat, lacks horns to meet the attack and may be fatally wounded.

The ritual battles of ibexes and bighorn sheep are stunning sights. The rivals rise up on their hind legs and hurl themselves with all their might at their opponents—who meet them with a similar charge. The clashes are repeated again and again, scores of times. Neither contestant tries to strike at his rival's body. All his force is aimed at and absorbed by the rival's strong horns.

These fights are not playacting: they are out-and-out displays of strength and agility at the end of which the contestants are exhausted. If an ibex wished to finish off his rival, as a predator does, he would have approached it stealthily and tried to gore its body with his horns. But an ibex or a gazelle initiating a ritual fight approaches slowly and shows off his intentions by lowering his horns. He waits for his rival to prepare, and if the rival does not, the challenger may butt it lightly to get its attention. By letting his rival prepare for combat, the ibex is taking on a handicap; the handicap pays off for the winner because he convinces his immediate rival as well as any onlookers of his superiority. This saves both animals the risk inherent in a battle to the death, and one or both combatants may demonstrate a prowess that saves it from having to deal with other rivals.

Contests of force don't necessarily involve physical contact. Gulls threaten by tearing up grass, chimpanzees break off branches, rhinoceroses and bulls rake up dirt with their hooves. Peacocks fan out their tails and hold them upright, shaking them, which demands a great deal of strength. Since males of these species perform the same actions, they and others can judge reliably which of the rivals is stronger, and thus the contestants are spared a fight.

· · ·

RITUALIZATION: DOES IT REDUCE THE INFORMATION CONVEYED?

Huxley,[3] who studied the courtship of the great crested grebe, suggested the term *ritualization* for the process by which movements that serve as signals are derived from movements that originally had some other function. The courting male grebe repeatedly touches the sides of his body with his beak, as if to preen his feathers, but he is not actually preening himself. When preening, a grebe combs each feather from base to tip. The courting movements are much more stylized; the same feathers are touched repeatedly but are not combed. This formalized movement reminded Huxley of the formalized gestures humans use when they perform rituals, hence the term ritualization.

Huxley suggested that ritualization evolved in order to differentiate between the original movement—preening—and the one used for signaling. He suggested that the most important function of ritualization is to increase the clarity of the signal, in order to differentiate between it and the original, functional movement, and between it and other signals. This opinion is still accepted by most researchers.[4]

Even though movement signals are uniform, they are not performed in exactly identical ways by different individuals, nor by the same individual in different circumstances. This difference is not accidental; it often shows the intensity of the signal. For example, if stretching the neck indicates fear, then the degree to which it is stretched indicates the degree of fear. Morris[5] suggested that movement signals evolve by two conflicting processes: the movement becomes more uniform so that the receiver of the signal will recognize the message, while variation emerges to show magnitude. He believed that these two competing tendencies led to a compromise that required movement signals to be more formalized than the movements from which they had developed but allowed for enough variability to show quantitative differences.

Morris's idea is logical, but it does not address reliability in movement signals, and it does not explain why certain movements rather than others were selected for a specific message. Both questions are answered by the Handicap Principle.[6] If movements that develop into signals must demand enough effort that cheaters cannot use them or will not find them worth their while, then individuals of different abilities and who are more or less motivated would perform them differently. The difference in performance reveals specific differences among performers, which observers are interested in. The selection of reliable signals by their receivers brings about ritualization, and the cost of performing the formalized movements results in variability; there is no need to posit two *conflicting* selective pressures. Rather, ritualization and variability are *complementary* results of signal selection.

Morris assumed that ritualization—the standardization of movements for the sake of clarity—reduces the information contained in the movement messages. But is there really a loss of information? We think not. In fact, it is precisely the need to compare two or more performers and see clearly the differences between or among them that leads to standardization.

While working on this chapter we happened to be watching TV coverage of the Olympic games in Barcelona. The degree of standardization achieved by sprinters was impressive: they started off at the same instant and were no more than hundredths of a second apart at the end. Athletic contests are governed by rules. All contestants perform the same tasks. Untutored viewers may have trouble deciding which of the several top-notch gymnasts is the best. After all, experienced judges focus on details that seem almost inconsequential to the uninitiated. For example, a great deal of a gymnast's success depends on the way he or she gets off the apparatus at the end of a performance: ideally, a gymnast lands with perfect balance, feet aligned and straight. This may seem to a lay observer to be an arbitrary, draconian caprice of the judges, but closer examination shows that the smooth, rapid transition from movement to a perfect stance at the end of a grueling effort is indeed one of the hardest parts of the exercise. When one has merely to tell a good performer from a bad one, judging is easy. Olympic judges set minutely exacting standards because they have to judge among many outstanding, top-notch performers.

Individuals performing movement signals in the animal world are like competitors in an athletic contest. The signalers show off their ability to execute a certain movement, while the observers are the judges evaluating the signalers' performance. The judge may be a female looking for the best father for her offspring; a predator looking for prey it can catch; a rival evaluating whether it has any chance against the performer; or a parent deciding whether its offspring should really get more food.  These "judges" force the competitors to compete in a standard manner in order to better evaluate the differences among them. It is precisely this standardization that brings out crucial differences in performance, which in turn reflect accurately the different abilities and motivation of the competitors.

The inexperienced may not notice subtle difference in performance and may think that all signalers are giving the same message in the same manner. This is not the case. Simpson remarks that experienced observers can predict the results of an ongoing fight between two Siamese fighting fish: the one that is holding its gill covers erect most of the time toward the end of the encounter is likely to be the winner.[7] This seemingly minute and trivial gesture, in the context of a fight, means that the fish is giving up the full use of its gills and is handicapping itself by absorbing less oxygen than it can; this is reliable evidence of its stamina and its chances of winning the fight. In nature, just as in the Olympics, seemingly negligible differences in performance can bring about dramatically different results. In the Olympics, the runner who comes in first gets a gold medal, while the one in fourth place gets no medal at all, even though the difference between them may be so small it can only be seen with the aid of electronic devices. In the wild, in species whose males do not participate in the rearing of young, a very small number of top performers get to father most of the next generation, while many males have no offspring at all.

HOW A RITUALIZED SIGNAL EVOLVES

How, then, does ritualization work? How do movement signals evolve from movements that used to have another purpose?

A signal's value to the signaler is that it can convey information to another individual. But the message can be conveyed only if the other individual is interested in the message and understands it. The process therefore cannot start with a mutation in the signaler; because that would require *two* simultaneous, coordinated mutations: one which caused the signaler to perform the signal, and another causing the observer to take interest in it and understand its meaning. Even in the highly unlikely event of two such simultaneous mutations, the chance of the two mutants meeting is practically zero.[8]

 But animals constantly collect information, including information about the behavior of such other individuals as offspring, mates, rivals, predators, and prey. One reason is that they need to be able to anticipate what other individuals will do. Most actions involve preparatory movements: a predator has to look and aim before striking; a grazing deer has to shift position before fleeing; a bird gathers itself to fly. Observers can often anticipate animals' actions. Cats stalking their prey and dogs preparing to charge watch their prey's activities. An experienced birder—or a predator—freezes when the bird stops feeding and looks at the observer: the fact that the bird has stopped feeding and is stretching its neck means it has noticed the observer and may fly away. Likewise, an animal that notices a predator staring at it intently can anticipate the attack and flee. The observed movements are not "signals"—they may convey information, but they are not performed for that purpose. Krebs and Dawkins[9] called this stage in the evolution of signals "mind reading."

Such preparatory movements are made regardless of whether or not anybody is watching them, because of their intrinsic usefulness, rather than to convey any message. The ability to watch for specific movements other than signals can then evolve and spread through a population in the same way that any other single, useful mutation does.

Once observers respond to specific observed movements, mutations that exaggerate these movements may spread among the ones observed. These mutations are an advantage to their bearers because they intensify the movement in the eyes of observers who *already understand its meaning*; at this point the movement has become a movement signal.

There is a cost involved: as soon as a movement becomes exaggerated, it ceases to be the optimal movement for its original purpose. Still, the investment is worthwhile to the signaler, who wants to make sure its actions are noticed. When a bird stretches its neck exaggeratedly to ensure that the predator—or birder—knows it has noticed it, it has to stop feeding for the moment and will have more difficulty in launching itself into flight, since in order to do so, it would have to pull its neck back. But the bird also lessens the likelihood that it *will* have to take flight and leave its feeding ground: the potential prey makes it clear to the watcher that the watcher has been spotted and that approaching would be a waste of time. Both bird and predator gain from the signal: the predator can save its energy for other, more vulnerable prey, and the bird does not have to leave a feeding or resting spot.

But why do movements that evolve into signals become exaggerated in one way rather than another? This is where the Handicap Principle applies. Let's return to the wolf–gazelle interaction that opened this book. Even before gazelles started signaling to wolves, individual gazelles reacted differently to a wolf's presence— they prepared to flee according to their confidence in their ability to evade the wolf. Weak individuals fled right away, while confident gazelles could afford to stand and watch the wolf, or even to continue feeding. In fact, Kruuk describes a cheetah bursting into a crowd of Thompson's gazelles and catching one of them. He stresses that the gazelle the cheetah caught was the one who first turned to flee when the cheetah began its charge.[10]

A wolf could learn by experience that it had a better chance of catching a gazelle who fled right away. It might even pay off for the wolf to advance more slowly toward the gazelles and let them react, so that it could spot and chase the weaker ones. A confident gazelle can make it easier for a wolf—and for itself—to avoid a futile chase by demonstratively postponing its escape, or by exaggerating movements that display its fitness and self-confidence, like jumping straight up (stotting). Individuals will perform the signal differently; their performances will differ even more than their nonsignal reactions to the wolf's presence, because the wolf pays most attention to those movements that most reliably demonstrate individuals' differing abilities to escape it. It is precisely these movements—the ones that prove to be reliable means of communication—that became formalized, ritualized movement signals.

Exaggerated movements used as signals are constantly tested by natural selection. The signal evolves and endures only as long as watchers gain from it reliable information that is beneficial to them, and as long as the investment in it is worthwhile to the signaler. The benefit in watching another to anticipate its actions is what generates movement signals; the specific character of the movement that evolves depends on its specific cost in the circumstances in which it is used. For example, movement toward a rival became a threat, even though sideways movement would have demanded as much physical effort and might have given the rival a better sense of the details of the movement. In this case, it is the risk taken by approaching the rival, rather than the physical difficulty of performing the movement or its specific details, that reliably displays the degree of threat.

The standardization and increased difficulty of movements used as signals—their ritualization—makes it easier for watchers to determine the differences among displaying individuals, and in the reactions of the same individual in different circumstances. If each signaler displayed in its own way, observers would find it difficult to compare them. Movement signals thus become ritualized not so they can be distinguished from the functional movements out of which they evolved, but rather to enable observers to make fine comparisons among signalers. The logic of movement signals is the same as that of species-specific patterns of color and form: *uniformity within a species*—ritualization—evolves out of the competition members of the species engage in to demonstrate their *differences*. In both cases, it is the observers, the ones the signals are directed at, who by their choices drive the evolution of reliable and uniform signals.[11]

CHAPTER 6

VOCALIZATIONS

M any animals make decisions based on information they get from the vocalizations of others—their mates, their collaborators, their rivals, members of other species. What is it that leads individuals to trust information they get from others' voices? And why isn't vocalization used to mislead hearers? It would seem easy to cheat by vocalization; yet observations and studies show that calls usually convey reliably the intentions of the callers. Dog and cat owners can often predict their pets' behavior by listening to them, and we don't know of any case in which a dog tried to mislead by using its voice.

THE CORRELATION OF VOICE WITH POSTURE AND TENSION

We discovered the key to the reliability of vocal signals one evening at home. We found out that we could not make a sigh of relaxation while getting up, nor could we make a convincingly frightened cry while settling down comfortably. To make a sound of fright, one has to tense one's muscles—which is impossible to do while relaxing in an easy chair. On the other hand, to make a sigh of relaxation one has to relax one's muscles, and that is impossible to do while getting up. This simple experiment made us realize that vocalizations might be faithful representations of the state of the body producing them.[1]

As we said in the discussion on rivals, actors can sound convincingly loving, hateful, threatening, or depressed without actually loving or hating their fellow actors or feeling despair. In order to do so, they have to assume the stance and movements typical of the feeling they want to convey. Opera singers, too, know that their movements on stage affect their voices and learn to move in ways that let them sound the way they want to.

If we assume that the body is the resonator, the sounding board, then it stands to reason that the quality of vocalizations is affected by the state of that body. Vocal pitch, too, reflects the tension of the muscles of the body and face.[2] This relationship between vocalizations and the body that produces them makes them hard to fake. To make a deceitful call, one has to adopt the posture necessary to produce the fake message; since each action has its own optimal starting position, changing position to produce a false message may make it extremely difficult to carry out the action really intended. Thus what one gains by cheating does not make up for what one loses by assuming improper posture—and so the phony message is not worth its price.

In a dolphin research center in Hawaii, dolphins were taught to make sounds according to the researchers' request. When the dolphins were asked to make a threatening call, they made a sound somewhat different from true threatening calls. When asked by their trainers to improve, they made a sound just like a real threat call—and also assumed the facial expression of a dolphin who is angry and ready to bite. They were not angry; their caretaker, who would not dream of putting his hand into an angry dolphin's mouth, could put his hand in their mouths with impunity. But they had to assume the *posture* of an angry dolphin in order to make the *call* of an angry dolphin.

Interestingly, Darwin observed that when we listen to singing and music we interpret the notes according to the actions of the muscles that produce them.[3] But Darwin was not concerned with reliability in communication. Others, too, have remarked on the connection between body state and voice: Scherer[4] pointed out that vocalizations express the motivation of their producers because their tonal quality is affected by the stance typical of that motivation.[5] But nobody saw in this connection a key to reliability in vocal communication in general.

• • •

THE INFORMATION CONVEYED BY VOCAL SIGNALS

One can often tell what others are doing merely by listening. This applies both to people and to animals. Rowell[6] found she could tell by listening what was going on in a confrontation between monkeys. In a radio program several years ago, it was said that in the newborn intensive care unit of the Soroka Hospital in Beer-sheba, Israel, the nurses could tell which part of a newborn baby's body was hurting by the sound of the baby's cry. Apparently, pains in different parts of the body make different muscles tighten, and these muscles affect the sound of the cry in a way that a trained observer can interpret.

The crucial importance of small differences in vocalizations as a source of information is brought out by an experiment that Gaioni and Evans conducted with ducklings.[7] When a duckling can't find its mother and siblings, it emits a series of peeps. When it hears another making peeps, it stops and listens. Individual peeps vary slightly, and the researchers—who wanted to study the most effective components in the peeping—compared a natural series of peeps with an idealized one, produced by computer, which repeated a single peep. To their surprise, ducklings quickly lost interest in the idealized series that was free of all distortion and variation. The natural series caused the ducklings to stop again and listen.

A lost duckling probably stops peeping in order to concentrate on listening so that it can get the most information from the peeping it hears. It is precisely the variations, the ongoing changes in its sibling's situation, that concern it: Has the sibling seen their mother and started running to her? Is it hesitating? Is it frightened? An artificial, uniform series of peeps reflects no change and therefore conveys little information that the duckling is interested in. The continued peeping conveys not merely the fact that a duckling is lost, or even which duckling is lost, but rather information about the ongoing changes in its circumstances and actions.

A similar logic governs the daily greetings we exchange with one another. Do our coworkers really need to be reassured that it is a good, rather than a bad, morning? The greeting enables our colleagues to detect, by the tone of our voice, fine differences in our daily moods. A colleague's mood is very important to people working with him or her—they need to know what to expect from their colleague that day. Since mood affects one's posture and body tension, we can tell the mood of our acquaintances by listening to the sound of their voices. It is precisely the standard nature of the greeting that lets us detect fine differences in mood—detection that is to the benefit of both greeter and listener. This simple method seems to be far more effective and reliable than would be long, detailed, wordy descriptions of our daily moods. And indeed, sometimes the response to a good-morning is, "Are you all right? What happened?"

• • •

ANIMAL VOCABULARIES: THE CONNECTION BETWEEN THE MESSAGE AND ITS VOCAL PATTERN

If any sound that an animal makes reflects the state of its body faithfully, why do animals use particular vocalizations to communicate? Clearly, any mood—fear, gladness, joy, sadness—will be reflected in any vocalization that we make. But fine differences in a particular mood—precisely the variations that the listener is interested in—seem to be displayed better by some vocalization patterns than by others. Different vocal patterns are affected differently by changes in posture and body tensions. It makes sense that the vocal pattern most suited to show fine gradations in anger is different from the one best adapted to show degrees of relaxation. Some vocalizations will show better the difference between sad and very sad; others will highlight the difference between simple gladness and supreme joy. A sigh of relaxation is best suited to show the difference between one level of relaxation and another; the sharp yell that sometimes goes with a jump is best suited to distinguish between different jumps.

It is commonly assumed that the vocal pattern typical of a mood is meant to convey the general mood of the vocalizer to listeners, that is, to say "I am sad," "I am angry," "I am pleased." It is indeed possible to compile "dictionaries" that describe the vocal pattern typical of each mood: joy, despondency, boredom, wakefulness, appeasement, aggressiveness.[8] But it does not follow that vocal patterns evolved to convey this dictionary-type information; circumstances are often sufficient to convey it without any vocalization at all. The information the listener lacks is the precise degree or nuance of the mood. A threatener, for example, needs to convey not the obvious fact that it is issuing a threat but rather the degree of that threat. The degree of threat is not known in advance and indeed may change from moment to moment. The vocal pattern best suited to show fine differences in the degree of threat is the one that is best adapted for threatening.

Obviously, once vocalizations optimal for showing off fine differences in specific moods have evolved, one can make secondary use of them to identify the type of mood. We know that crying usually indicates sadness or pain; laughter, a happy mood; screaming, anger or alarm. But we believe that this function played no part in the evolution of vocal patterns. It is parallel, in our opinion, to the use of species-specific patterns to identify members of a group, after those patterns have evolved to show off differences among individuals *within* the group.

RHYTHM

As we saw in chapter 3, Lambrechts and Dhondt[9] found that the more successful great tits—the ones who produced more offspring—had songs that contained more syllables and were more rhythmical than the songs of other great tits. Among humans, more precise rhythm in threats tends to inspire greater fear; in movies and on stage, rhythmic music can announce important or fearful happenings.

What is the connection between rhythm and threat? There is an inherent conflict between collecting information and precise execution of vocalization. Both activities demand concentration, but to collect information one must concentrate on listening, looking, discerning, and correlating, while precise vocalization demands that one concentrate on execution. A person who tries hard to catch a weak or unclear sound often closes his or her eyes. And many musicians close their eyes, or assume a dreamy expression, when concentrating on their playing; they disconnect from their surroundings.

One who tries both to listen and to vocalize is likely to falter a bit in the rhythm—a stumble that will display divided concentration. Therefore, a precise rhythm shows that one is not collecting information. One who is certain of his or her ability to win or is determined not to give in does not have to collect additional information. But one less confident, who is still trying to decide which way to act, needs any information available. This difference between the confident and the wavering is brought out reliably by a rhythmic threat.[10]

Anava studied the trilling calls made by babblers when they mob perched raptors.[11] Each trill is composed of a rhythmic series of short notes and pauses. A babbler mobbing without interference from others tends to make loud, prolonged, uniform trills composed of many syllables. When the mobbing babbler is disturbed by other babblers, however, its calls are affected. When a dominant bird approaches the calling bird, the caller's rhythm is appreciably less even. It is also altered, though less so, when the mobbing babbler approaches group members who are subordinate to it. The difference is both audible to humans and clearly visible in recorded analyses (see Graph 6-1). Interestingly, when babblers get nearest the perched raptor during their mobbing, they fall silent. Apparently, assuming the mobbing posture of raised wings and spread tail interferes with the babbler's ability to escape swiftly should the perched raptor try to strike. Thus, the mobbing babbler cannot afford to take some of its attention off the raptor in order to perform vocalizations. Although those mobbing babblers who come nearest snakes adopt similar postures, they make short *tzwick* calls; it would seem that snakes are somewhat less dangerous to babblers than perched raptors.

VOCAL PATTERNS USED OVER DISTANCE

One of the most important factors affecting vocal patterns is distance. Babblers standing next to each other usually communicate with one-syllable vocalizations, while their communication over greater distances takes the form of loud, multisyllable calls. Physical barriers—trees, rocks, water, fog—distort fine vocal pat-

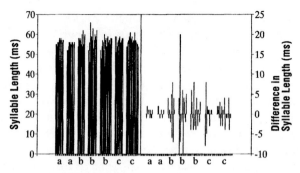

Graph 6-1. Mobbing trills of the male babbler TTZL when alone (a), when disturbed by the dominant male of his group (b), and when disturbing a subordinate (c). Data from Anava, 1992. The left side of the graph shows the length of each syllable within each trill; the right side shows the change in length between each syllable and the one before it. The difference between states a, b and c may seem insignificant on the left graph, but when one focuses on the differences, as in the right graph, they stand out distinctly. Again, these differences are clearly audible to the human ear—not to mention to other babblers.

terns; so does the need to make a call loud enough to be heard over great distances. The differences that high volume does not convey well are often the very ones that the call is meant to communicate. For that reason, vocal communication over distance usually consists of a series of syllables, and rhythm and changes of volume convey a great deal of the information. Such a series also gives the listener time to locate the source of the sound, turn to it, and thus better perceive fine variations. Indeed, Katsir showed that babblers' calls vary according to the distance they are meant to cover: the greater the distance, the more constant the frequency and the greater the variations in syllable duration.[12]

WHY SHOUT?

It isn't always easy to tell just whom a vocal communication is directed at. As we noted already, the discrepancy between the loud "warning calls" a sentinel babbler makes and the short distance between the sentinel and other members of the group led us to understand that the calls were actually aimed at the distant predator. But sometimes babblers make loud calls even when clearly communicating with others nearby. Humans, too, sometimes shout at nearby listeners.

Everyone knows that there is often a correlation between volume and the degree of anger or of a threat—the louder the shout, the more intense the anger or threat. We suggest[13] that in these cases too the shouting is aimed not at the purported listener but rather at other babblers—or humans—who are farther away and are not parties to the conflict. The shouting makes them witnesses. When someone threatens another in private and then does not carry out the threat, he or she loses standing in the eyes of that person only. When others are made witnesses, failure to carry out the threat will cause the threatener to lose standing not only in the eyes of the one threatened but also in the eyes of the witnesses. By shouting, the threatener raises the stakes and makes the threat more reliable; only confident individuals can afford to shout their threats before the crowd. This principle is well-known in politics: a publicly declared intention is likelier to be carried out than one agreed upon in secret.

THE DURATION OF VOCALIZATION: REQUESTS AND COMMANDS

The duration of vocalization, like its volume or rhythm, is correlated to its message. The more persuasive a requester feels compelled to be, the greater the duration of his or her request, in any language: "Please"; "Could you please"; "If you really don't mind, could you please" and so on. Likewise, the more nestlings repeat their begging, the more urgent their parents understand the begging to be. But what has duration to do with requesting?

The point of making a request is to get something that is of value to or demands an effort from the giver. After all, if the thing were valueless to its owner, it could be taken without any fuss. Logically, the giver will consent to give up the valued object or spend the effort only if it either cares about the petitioner or gains from the act of giving. The potential giver already knows how it feels about the requester and about the item asked for; what the giver does not know and needs to find out is the importance of the requested object or action to the requester. The more highly the petitioner values the item or action, the more the giver gains by handing it over.

The increasing length of the request convinces the giver that the petitioner really needs and values the object or action in question, because the act of petitioning is costly to the requester. It demands energy and time and exposes the petitioner to predators or rivals. It also decreases the standing or prestige of the requester, and increases the standing of the giver, in the eyes of witnesses—who are more likely to notice a longer request, and to note its significance. The longer the petition, the higher the price the petitioner has paid for the item it wants.

A command, on the other hand, is best kept short. A short command that achieves its objective means that the one ordered was attentive to and ready to obey the commander. Thus a short command that is obeyed raises the prestige of the one issuing it.

DIALOGUES AND THEIR SIGNIFICANCE

A dialogue can reflect reliably the relationship between those who participate in it. It has been shown that a male bird interrupting another's song with its own is threatening the other.[14] One who waits for another to finish before itself starting to sing demonstrates either submission or a willingness to cooperate. The reason, again, seems to be the inherent conflict between vocalizing and listening. By not waiting for another to finish, a bird shows that it does not need the information the other provides in order to plan its action. But one who is unable to confront the other or one who would rather not do so and is willing to collaborate, has to pay attention to any information the other offers, because it may be essential if a conflict is to be avoided. Hultsch and Todt[15] remark that nightingales can mimic a neighbor's song either in goodwill or in rivalry. When the interaction is friendly, the nightingale starts singing, mimicking its neighbor's song, in the breaks between its neighbor's choruses; to signal rivalry, it bursts into imitative singing while its rival is still singing. Very old males do not respond to their neighbors' singing—possibly showing the indifference that attends very superior status.

Another way to show attention is the complex "duets" sung by many songbirds in the tropics. The duet is so well coordinated that listeners find it hard to imagine that it is sung by two birds rather than one. The duet has a fixed pattern. When one mate starts singing, the other has to drop instantly whatever it is doing and join in. This readiness shows off interest in the partner and the willingness to invest in the partnership. Laughing thrushes, group-living birds related to the babblers, demonstrate the cohesiveness of their group by coordinated group singing on the borders between their territories and those of other groups. Group singing by humans, too, often serves to demonstrate cohesiveness and care for the group—be it by worshippers in church, a student choir, children in summer camp, soldiers on the march, or just people getting together in order to sing.

MIMICRY

It is fairly common for animals and especially for birds to mimic the calls of other individuals or other species and even background sounds. Payne[16] and McGregor[17] found that buntings and tits mimic the songs of their neighbors; mimicking by singing crested larks in Hatzeva have often caused us to stop and look for babblers, and a friend of ours often ran needlessly to the phone, deceived by the mimicking of a jay. The top marks in mimicking belong to parrots and mynas, who can mimic human voices and use words and sentences in appropriate circumstances.[18] It would be fascinating to study the use and value of the imitative faculties of parrots and mynas in the wild.

It seems that mimicry fulfills several functions. One is to convince a listener that the communication is addressed to it specifically, as we saw in chapter 1. Ofer Hochberg[19] told us of a jay near his house that would call like a cat when it saw a cat. Alan Kemp of South Africa several times observed a drongo that mobbed a Wahlberg's eagle while uttering mimicries of the eagle's calls mixed in with its own alarm calls. The drongo near Kemp's home knew the calls of several hawk species and would mimic them correctly when they passed by.[20] It may be that both drongo and jay were trying to convince these predators that they had seen them specifically.

Birdsong, among its other functions, proclaims the singer's readiness to defend its territory. By mimicking a neighbor's song, a singer addresses its threats to that neighbor in particular. Tits, nightingales, and the males of many other territorial bird species often mimic the rival they are addressing.[21] As they grow older, males of many species, such as the lyrebird, the bowerbird,[22] and the nightingale, expand their repertoire of mimicked sounds and of songs. It may well be that their ever more variable mimicry lets them show off their experience, their age, and their ability to learn and remember. Humans, too, mimic sounds extensively, and Darwin saw in this ability the foundation of the ability to speak.[23] We believe that mimicry still plays an important role in human linguistic changes such as shifts in vowel sounds, the adoption of new words and phrases, changes in the meaning of existing words, and the like.

DO ANIMALS HAVE VERBAL LANGUAGE?

There have recently been successful experiments in which researchers taught apes to use some sign language and parrots to use words. Researchers next tried to interpret animals' calls as words. The best-known of these attempts is a study of the vervet monkeys in Amboseli in Kenya.[24] It turned out that the vervet monkeys used one call when eagles approached, another for leopards, and yet another for snakes. The researchers also found that these calls caused reactions in others that were appropriate to the particular predator, even when the calls were sounded by a tape recorder.

As we said in chapter 1, we think that the "warning" calls of babblers are aimed not at other members of the group but at the predator. Do the monkeys' reactions

to the calls prove that these calls are meant to warn? And can one compare the calls to words? Words—vocal patterns that serve as *symbols* representing specific meanings—have no inherent *connection* to those meanings; the symbols are arbitrary. But we believe that all non-human animal vocalizations demand a logical connection between the vocal signal and its message.

Babblers, too, make different calls in response to different predators. In the case of babblers, the calls seem to bear a direct relationship to the state of the body of the babbler making them. Rather than being symbols for a particular predator, the calls seem to be reflections of different physical reactions, each appropriate for a specific type of predator or set of circumstances.

When a raptor appears in the distance, babblers sound loud, abrupt calls that are like barks. The barking babbler is not afraid; it usually goes to the top of the canopy and follows the raptor with its eyes. To make the loud barking sound, the babbler has to expel a lot of air from its lungs. Trying to make similar sounds with our own bodies, we found we had to relax our muscles and could not repeat the barks without refilling our lungs with air between calls. If the bark reflects a similar relaxation in a babbler's body—which would make sense—it would seem to empty the lungs of air, which would handicap a babbler attempting to dart aside or into shelter. The call also discloses the babbler's location. In other words, a bark reliably shows the babbler's disdain for the raptor and its confidence that it can escape if attacked.

When a babbler sees a raptor diving at it, its call is entirely different: it darts into the bushes with a loud squeak. This "fear cry" reflects faithfully the tension of muscles used in the act of escape. This information may well deter the raptor from further attempts at that particular bird by showing off how quickly and effectively the later could dart into shelter.

When a raptor lands on a nearby bush, perhaps in the hope that the babblers will become less alert in time, the babblers sound a loud trill like a cicada's. This sound may last for tens of seconds, and repeated series of such trills may follow one another for a considerable time. The trill is a reliable way for the babblers to signal that they are still actively watching the raptor—since one who is relaxed and inattentive cannot make a rhythmic trill. On the other hand, the trill does not overly endanger the triller, since a continuous sound like a trill is difficult to locate, unlike a bark, which pinpoints the caller's location immediately.[25] As if to hammer home the point, whenever the raptor moves, the babblers react with a sharp, tense-muscled tzwick, which tells it yet again that they are following it constantly.

Babblers mobbing a snake make short, sharp sounds that reflect the considerable tension of their bodies. From time to time one of the babblers spreads its wings upward, and at that moment the sound it makes is somewhat different. These sounds are usually called "snake warning calls," but if they are warnings, why do the babblers come nearer the snake and raise their wings? We think that by approaching a snake, babblers show off their courage to other members of their group[26]; the change in their posture necessarily affects the character of their calls. Babblers who see a predator on the ground—a wolf or a cat—make a series of short, rhythmic tzwicks, which accompany and

reflect their jumps from tree to tree as they follow the predator. These calls, too, are faithful representations of the babblers' movements.

In short, babblers, like vervet monkeys, make specific calls in response to specific enemies. It is easy to show a correlation between the type of predator and its behavior and the chance that a babbler who sees it will make one call rather than another: a babbler will usually bark at the sight of a raptor in the distance, utter a sharp squeak at a raptor diving at it, trill at a raptor sitting on a tree and at a terrestrial predator in the distance, make "snake calls" at a snake, and emit tzwicks at a predator walking on the ground nearby. Like the monkeys, other babblers react appropriately to these calls: they join the sentinel at the top of the tree when they hear a bark, dart in panic into the brush when they hear a squeak, and gather to mob snakes or perching raptors. But we don't think that the various calls are meant to announce the nature of the danger to other babblers. The calls are directed at the predator, and the other babblers merely know from experience which circumstances each call accompanies.

Interestingly, babblers sometimes seem to "err" in their calls: they may bark at an approaching ground predator, or issue snake calls at a raptor or a human hand approaching their nest—at which time they also raise their wings, as they do when mobbing snakes. It is unlikely that the babblers actually mistake the raptor for a snake, however. Rather, in these particular circumstances it seems that responses other than the usual ones are appropriate to the specific situation, and that the calls reflect not the type of enemy but rather the particular motions of the babblers in reaction to the enemy. In each case, the call is appropriate to—in fact, is a product of—the movement. The calls give a faithful vocal representation of the babbler's doings from moment to moment.

Researchers report that vervet monkeys, too, sometimes "mistakenly" use an "inappropriate" call in reaction to a specific enemy. One wonders what the circumstances were in those cases, and what the monkeys would "say" if presented with dangers in unfamiliar forms—for example, an eagle in a cage on the ground. The assumption that the monkeys label their predators seems based on the reactions of other monkeys who hear the call, but those reactions do not prove that the callers intend to let others know what the specific danger is. Of course, others are very likely to be aware of the connection between specific calls and specific dangers: many animals react appropriately even to the calls of other species, which are clearly not directed at them. But we do not think that such calls evolved *for the purposes* of letting others know the enemy's identity.

It is well known that animals can correlate things with specific sounds, including words. This ability is used by trainers who teach animals to perform in response to commands—to words. The trainer teaches the animal to understand fragments of his or her word language. Dogs, dolphins, and parrots can learn persons' names and can on command sit, stand, turn left or right, fetch specific items and put them in specific places. Some animals can understand simple concepts encoded in words, and can make specific sounds on command. Why, then, did they themselves not evolve a language of words, as humans did, to describe to their partners where exactly food is located, to alert them to anticipated dangers that are not yet present, or to convey any other information that is best expressed by words?

In the final chapter we will discuss the special nature of human language, its strong points, and its drawbacks. Here we shall only say that, with all its potency, human language has no component that guarantees its reliability and prevents cheating; nor is it in and of itself well suited to convey exact, fine gradations of feeling and intention. Thus, in no human society does the language of words replace wordless communication—communication by nonverbal sound, intonation, stance, and movements. And this wordless, nonverbal communication is the only "speech" of animals.

CHAPTER 7

BODY PARTS THAT SERVE AS SIGNALS

The combs of jungle fowl are a good example of body parts that serve purely as signals. They have no other practical purpose. The comb is bigger in roosters than in hens, in older males than in younger ones, and usually bigger in dominant males than in subordinate ones.[1] The comb is a delicate, unprotected organ; it is full of blood vessels—which, as we have seen, advertise the health of the bird. No wonder that cocks, when they fight, try to injure their rivals' combs: a profusely bleeding comb may well determine the fight.

In the wild, a large, intact comb on a cock's head is evidence that in spite of such a handicap, no rival has yet managed to injure him: the comb's owner is clearly able to overcome rivals without being injured himself. As a result, a cock with a bigger comb can often deter rivals without fighting, since his large, intact comb proves his success in previous battles. Holder and Montgomerie found that ptarmigan males attack each other's combs; they also found a correlation between the state of males' combs and the number of females who chose to breed with them.[2] In the past, people who raised fighting cocks used to cut off the cocks' combs and dangling facial flesh:[3] fighting cocks raised by humans were supposed to kill their rivals, not deter them with effective threats. A cock without a comb may be able to fight better, but as it cannot threaten reliably, it cannot win without a battle.

LONG TAILS: ARE THEY SIGNALS?

When a body member has both signal and nonsignal functions, its shape is affected by both. For example, the peacock's tail is clearly a handicap and thus a signal. The impressive long train evolved from back feathers[4] and hampers the peacock both when it walks and in flight. The smaller supporting fan of tail feathers holds up the decorative back feathers. But peacocks still use their tails as rudders as well—though not the most effective rudders.

In theory, there is a clear difference between a signal and a feature that is not a signal: one evolves because it enables the animal to convey information and convince others that this information is valid, while the other evolves because of a straightforward utilitarian benefit to the animal. But in practice, it is sometimes difficult to tell whether a particular feature is a signal. The long tail of an adult male bird is not necessarily a signal; such a tail may evolve simply because it offers an advantage in flight to stronger individuals. The longbow of the Middle Ages enabled bowmen to shoot over long distances with great precision—but it required brawny, well-muscled arms. The bow was used because it was an effective weapon, not to prove its user's strength—but clearly, it could also serve the latter purpose.

Females may well look for longer tails because they are evidence of a male's strength, even though a longer tail that is actually useful to stronger males evolves under the selective pressure of the direct benefits it confers, rather than as a signal. After all, as Fisher remarked, observation by others of features that exist for util-itarian reasons precedes the evolution of the same features into signals. In other cases, the long tails are clearly handicaps: Evans and Thomas[5] found recently that shortening the tail of the scarlet-tufted malachite sunbird helped the sunbirds in flight. Evans[6] also found that the length of the tail depended on conditions during its development—which is what we have come to expect with handicaps.

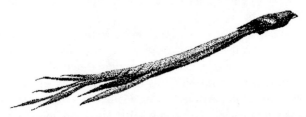

The tails of many birds, such as swallows and terns, have extralong feathers at the edges (outer tail feathers); in such birds as bee-eaters, yaegers, and tropicbirds these display feathers are in the center of the tail. They are usually longer in males than in females, and longer in older males than in younger ones. For the mating season, the male widowbird, a sparrow-size songbird of the savannas of East Africa, grows tail feathers a foot and a half long.[7] A widowbird displaying in flight above its singing post looks like a long, black, waving ribbon with a small bulge at the front—the body of the bird. Male widowbirds, like peacocks, have to bear such extreme handicaps because they do not participate in rearing the young: their only contribution to the next generation is their sperm. Thus, females do not mind sharing them with others and can afford to search for the few most superior males, who end up fathering most of the next generation.

BRISTLING HAIR OR FEATHERS: AN ILLUSION OF SIZE, OR A HANDICAP?

In egrets, birds of paradise, and other birds, some of the feathers of the back (the mantle) have evolved into long, threadlike decorative feathers. Most of the time these feathers are hidden among other, more utilitarian feathers; but on occasion they are held up and displayed to mates or rivals, framing the bird. It seems that the actual cost of developing and carrying such feathers is small; but there is a handicap involved in *displaying* these special feathers. They probably function like the long hairs on hyenas' backs or the long, soft quills on porcupines' backs, which stand up in a situation when the animal is making a threat or is excited. In order to make the feathers, hair, or quills stand up, the animal must tense certain muscles that may well interfere with attack or flight, which makes its threats more reliable.

The usual explanation for such bristling hair or feathers is that they increase the apparent size of the animal.[8] But an animal's body, or its head, in fact looks smaller within a frame of bristling hair than it does without such a frame. We are familiar with this effect in everyday life: the head of a person who is bald or 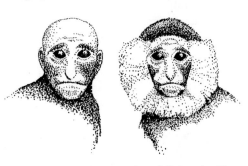 who shaves his head looks larger than that of a person with a full head of hair. Similarly, an egret bristling its threadlike feathers, a porcupine its long quills, or a hyena its hair makes itself appear smaller than it actually is. Animals display bristling hair or feathers not to *fake* size but to *demonstrate* it reliably; only a large individual can afford to make itself look smaller in the eyes of rivals or collaborators.

MANES AND CRESTS

The conventional explanation for manes, as for bristling hair or feathers, is that they deceive rivals about the size of an animal's head. If manes were meant to convince rivals that an animal's head is larger than it really is, however, one would expect them to be the same color as the head.[9] Yet that is rarely the case. The langur monkey has a golden mane and a black face. One can easily distinguish between the skull, with its jaws and jaw muscles, which are the langur's weapons, and the mane, which reflects the social standing of the male. Then too, if the purpose of the mane is to enable the animal to cheat, why don't youngsters cheat? After all, they would gain more from pretending to be larger than full-grown adults would. If, on the other hand, the mane is actually a handicap, only monkeys with larger skulls could afford to make their heads look smaller in this way.[10]

Crests of feathers similarly reduce the appearance of the beak, especially if the crest stands up, like a hoopoe's. The bill is an important weapon for birds, and a larger bill is a greater threat than a smaller one. Again, only a bird with a large beak can afford a large crest that makes its beak look significantly smaller. Both manes and crests, however, present other handicaps as well: crests make the direction of a bird's gaze evident, and manes interfere with vision.

HANDICAPS THAT INTERFERE WITH VISION

Sight is the most important sense in many species, and anything that interferes with an animal's ability to see is a very serious handicap. Compare, for example, the field of vision of a young orangutan with that of an adult male: the head of the former resembles that of a human child, and the youngster has a wide field of vision. Adult males have sunken eyes and a fleshy frame around their faces that impairs their peripheral vision; the older the male, the narrower his field of vision. Only a large male can afford to advertise his social standing with a handicap that

prevents him from collecting information from all around him. Such a handicap is too much for a young male. The manes of adult lions and of male langurs, baboons, and other monkeys restrict vision as well.

In our discussion of sexual display, we described the bulge that the courting white pelican develops on the base of its beak just in front of its eyes. The bulge blocks the pelican's view of fish near the tip of its beak, proving to potential mates that it is an excellent enough fisher to succeed even when thus handicapped; the bulge disappears by the time the pelican has to fish for its young. One can also find bulges that interfere with vision in some plovers and in Cracidae, large birds that gather fruits from the tops of tropical trees. Asiatic hornbills live at the top of tropical forests. Their gigantic bills are adapted for picking fruit from the end of branches. But some have large horny casks on their beaks, almost like another beak, which overhang their eyes. The cask impairs their ability to see birds of prey flying above and their ability to peck efficiently at fruit right in front of them, and also adds unnecessary weight to the beak.

Bustards are large fowl of the open desert and savanna. The neck of the male is covered with long feathers; when he displays, he stretches his neck far backward, and his neck feathers stand up, forming a large ball. He cannot see what is around him and dances blind. Only a male secure in his knowledge of his surroundings and confident in his ability to overcome predators and rivals alike can afford to dance for minutes at a time without being able to see, in full view of all comers. In fact, the bustard's blind dancing might be too great a handicap for it to bear in environments other than the open flatlands in which it lives; the specimen at the Tel Aviv University zoo repeatedly bumped into its fence while displaying.

BODY PARTS THAT EMPHASIZE THE DIRECTION OF GAZE

Another aspect of vision that signals often show is the direction a bird or animal is looking in. The California quail has a long black crest that curls forward. We were puzzled by it: the quail we were familiar with, the European quail, spends practically all its life on the ground in dense vegetation. What could such a bird show off with a crest of feathers? The mystery was solved when we saw the California quail in its natural habitat. Unlike the European quail, the California quail gives its loud call while standing on such high observation posts as the tops of rocks or bushes. Its black crest is visible from afar and shows the direction of its gaze. When the crest appears as a straight line extending upward from its head, the quail is gazing directly at an observer; when the crest appears arched, the quail is looking to the side. But by the same means, an observer can tell where the quail is *not* looking. The shape of its crest allows the quail to display its confidence; plumes on guardsmen's helms and Roman centurions' helmets may have evolved to fulfill the same function.

In Israel, there is only one lark that has a crest; the crested lark is also the only lark in Israel that sings standing on a hillock, a post, a fence, or a bush, as well as in flight. The crest shows clearly what direction it is facing; in flight, a crest is not seen and is not an effective showing-off signal. In humans and in some monkeys, noses show the direction of an individual's gaze. The nose is more developed in males than in females, and more in adults than in the young—an indication of its use for showing off. Small tufts of feathers on both sides of the head, as in owls and other birds, and short horns, such as the bony protrusions on the heads of giraffes, also let an observer see from afar which way the animal's head is turned.

The arched horns of gazelles, antelopes, and other such animals also show off the direction of their gaze; we found that out at the Shushlui Nature Reserve in South Africa. Till then we had always toured African reserves by car. At the Shushlui, a student doing research offered Amotz a chance to sightsee on foot. Unprotected and unarmed among lions, rhinoceroses, and buffaloes, Amotz couldn't help but listen to every sound and keep a constant watch on his surroundings. The animals were not used to the sight of humans walking; antelopes barked at the sight. Because of the distance, neither their eyes nor even the markings on their faces could be seen, but their horns always looked even and distinct to Amotz and his guide and followed them around, proof that the antelopes were looking at them.

The real adventure of the tour came in an encounter with a square-lipped rhinoceros. Square-lipped rhinos have two horns: the large frontal horn serves as a weapon; Amotz had a hard time trying to figure out the function of the second smaller horn behind it—until the rhino cow charged.

The guide had assured Amotz that square-lipped rhinos are usually peaceful, and that even if they charge, one can usually deter them at the last minute, or escape behind a tree. They came within a hundred yards of a rhino cow and her young calf. The guide, who probably wanted to show off, suggested getting nearer. Amotz had no choice but to come along; being left on his own was an even more frightening prospect than the rhinos.

The rhinoceros cow snorted, stamped, and charged. Amotz and the guide clapped their hands to deter her, and indeed at about twenty yards she swerved aside. The instant she turned, the small horn could be seen: until then it had been hidden behind the large, threatening frontal horn. It was only then that it became clear that the function of the rhino's smaller horn is to show what direction its owner is facing.

The small horn lets one tell whether a rhinoceros is confident or hesitating. A rhino with a well-developed back horn cannot hide its intentions; any hesitation, any change of direction or glance sideways is unmistakable. Such horns function this way only in an open landscape, where they can be seen from afar. The horns of Asiatic rhinos, which live in dense vegetation, do not show the direction of the rhino's gaze in a similar manner, and indeed their back horns have degenerated into small horny bumps.

BODY PARTS THAT HANDICAP FIGHTING

The special snout or proboscis of the elephant seal hangs down and covers its face. The snout prevents mature males from seeing objects right in front of their mouths, and they cannot feed or bite rivals without swinging it aside. The elephant seal's snout, like a cock's comb, can be grabbed and torn by a rival. The larger the snout, the better proof it is of the male's status: only a large, strong, and experienced male can afford to fight burdened with such a snout.

The puffin and a few other members of the Alcidae family nest in burrows; these birds develop a special horny layer around their beaks during breeding season. Richard Wagner, who studied the razorbill—a member of this family—found that the resident of the burrow protects it from within by presenting its beak to a rival in such a way that the intruder can grasp it.[11] The depressions in the rough, horny sheath of the beak let rivals get a good grip on it. If the beaks were smooth, as they are outside breeding season, rivals could not get a good hold on the beak of a puffin inside its burrow. Only a bird sure of its strength can afford to let its rival grab it effectively. The size of the beak's horny sheath, the number of grooves in it, and the decorations that show it off to best advantage all increase with age. Members of the Alcidae family who do not nest in burrows do not develop such sheaths.

The beard of the male ibex puzzled us for many years. Clearly it is a signal: adult males have them, but not females or young. But what does the beard show off? It is not heavy; it may change somewhat the appearance of its owner's head but is dwarfed in this respect by the adult male's gigantic horns. The solution came from a totally unexpected direction. Giora Ilani, who studied leopards at the Ein Gedi Nature Reserve, found that ibexes are an important part of the leopard's diet. A leopard can overcome even a large ibex: it grabs the ibex by its face, covers the ibex's nose and mouth with its own mouth, and suffocates it.[12] Anything, such as a beard, that makes it easier for a leopard to grab hold of the ibex's face makes it more difficult for the ibex to escape. An ibex that grows a beard is showing off its contempt for leopards and its confidence in its ability to escape them.[13] Another such body part is the loose skin on the lower jaws of adult male moose; wolves go for this dangling skin when they attack moose.

The heavy branched antlers of deer and the heavy arched or curling horns of goats and ibexes are not efficient fighting weapons. In fact, Darling[14] notes that the lunge of an antlerless deer is far more dangerous than that of an antlered deer. Yet it is clear that the larger and more branched a deer's antlers, the more attractive they are to females and the more of a deterrent to other males.[15]

From time to time one finds deer without antlers (hummel) or with straight antlers (pronghorns). Darling remarks that in battle, straight, sharp antlers can wreak a great deal of harm. But such deer disappear from natural populations: the straight antlers may be better weapons but are less effective when used to threaten and to display one's strength. It may well be, then, that heavy, cumbersome horns and antlers evolved not simply as weapons, but rather as handicaps that show off the strength of the animals that carry them. Females seem to be looking for males that are stronger overall, not for males who can kill better.

CAN BODY PARTS EVOLVE TO REDUCE THE COST OF SIGNALS?

The Handicap Principle states that every signal imposes a cost on the signaler— and this is true of the body parts we have just discussed. But can body parts evolve to *reduce* the cost of signals? During breeding season male skylarks perform as many as ten singing flights per hour. Moller[16] found that both the number and the duration of these flights vary greatly; a flight can take from half a minute to twenty minutes. He also found a reverse correlation between wingload—the ratio of weight to wing area—and the duration of each bird's singing flights. Moller cut off some of the wing feathers of certain males, reducing their wing area by 20 percent and increasing their wingload markedly. He found that such males made fewer and shorter display flights. He therefore suggested that male skylarks evolved larger wings that lessened their wingloads as "cost-reducing" features, to make display flights easier.

The question is, are the large wings and smaller wingloads actually an advantage to males otherwise, such as when they are escaping raptors, or during migration? If so, then they are the equivalent of the longbow—good for their own sake, and also good indicators of overall strength. In that case, the singing flights show off the larger wings to advantage: after all, according to the Handicap Principle, signals—such as display flights—are more affordable to stronger individuals than to weaker ones.

If, on the other hand, the larger wings are an advantage in display flights only and are a handicap otherwise, then both the wings and the singing flights are signals. In that case, the cost of the display flights would consist not only of the bird's investment in the flights themselves, but its investment in growing the larger wings and developing the strength to use them, and the liability larger wings pose in regular, nondisplay flight. In either case, cutting off some of the wing feathers surely affects the displays flights but proves nothing about the purpose of the larger wings of male skylarks.

THE EVOLUTION OF HORNS AND ANTLERS

The Handicap Principle can explain one of the dilemmas of evolution: the first stages in the evolution of new body members, such as horns, antlers or feathers. The laws of natural selection assert that only a mutation that benefits the individual who carries it will spread through the population. But how can we explain the chain of events that led to the appearance of horns on a previously hornless ungulate? It doesn't seem likely that a single mutation could create a pair of sharp, hard, symmetrical horns that serve their bearer efficiently.

The first stage in the evolution of horns and antlers may have been the appearance of small, soft bulges, like the present-day giraffe's, which served as signals showing the direction of an animal's gaze. This explains the location of horns on the forehead, as well as their general shape and symmetry. At this early stage horns would have been useless in a fight—in fact, they would have been a hindrance. Female Grimm's duikers have no horns; they fight each other by butting their heads. The males have small, slender horns and fight their rivals by kicking with their front legs. If they tried to butt their heads, their dainty horns would break. Such delicate horns are clearly not weapons but rather a signal showing off the male's quality. Once horns attained their place and general shape as signals showing the direction of gaze, they could then evolve into efficient fighting tools by means of mutations that improved their strength and sharpness (utilitarian selection). For example, the straight, hard, sharp horns of the females of several gazelle species are a useful, effective weapon against small predators.

Bigger, stronger males can use stronger, bigger horns, and that is just what utilitarian natural selection produced in the expected way. Then signal selection took over again, and made horns heavier and more elaborately spiraling, curled, or branching in order to show off such males' strength and stamina. The result was the heavy, back-curved horns of the male ibex, the spiraling horns of the kudu, the branching antlers of male deer. Such heavier, convoluted horns and antlers are

better indicators of each male's overall quality, and better instruments for ritual fighting, even though they are less effective as conventional weapons.

Changes occurring in this later stage of signal selection were confined to males: Females, as we have seen, have to invest more in each of their offspring than do males, since eggs are larger and fewer than sperm; female mammals have to invest even more in pregnancy and suckling. Thus they have less to invest in showing off. They also have less to gain: the number of offspring each female can bear in her lifetime is far more limited, whereas the number of offspring a male has depends far more on the number of partners who mate with him. Thus females have to go for quality and concentrate on carefully choosing the fathers of their offspring. Of course, this means that males are much likelier than females to end up with no offspring at all. Thus males show off considerably more than females do, particularly when the females do not need their help in raising offspring. This explains the great size and weight of male horns, and the wealth of complex shapes they take; by contrast, females, when horned at all, tend to have small, sharp, efficient horns for protection against predators.

But a further interesting stage in the evolution of antlers occurred in reindeer: their heavy, branching antlers evolved an extra branch, which turned out to be a useful tool for clearing snow off the lichens they eat in winter. This is probably the reason female reindeer—unlike other female deer—grow antlers.

The evolution of horns and antlers thus consisted of several alternating stages of natural selection. Originally, structures that signaled the direction of an animal's gaze evolved by signal selection, and their shape and position was determined by this function. These structures became progressively stronger, evolving by utilitarian selection into weapons. They continued to change, becoming larger and more elaborate as they evolved into handicaps and instruments of ritual rather than actual combat. And with reindeer, antlers further developed into food-finding tools. Still, throughout the saga, in all their various permutations, horns and antlers continued to serve as signals that showed the direction of an animal's gaze.

SIGNAL SELECTION AND THE EVOLUTION OF FEATHERS

Feathers, like horns and antlers, present a puzzle to researchers of evolution. Birds clearly evolved from reptiles, and feathers from scales. But it is unlikely that one mutation created something as complex and as beautifully functional as a feather. Nevertheless, in order for any change to spread in a population, it must be an improvement; one can readily see how a body member that has a particular function or purpose can evolve to better serve that purpose, but how can a body member like a reptilian scale evolve gradually into a totally different one like a feather? The transformation of a scale into a feather can only have occurred through a series of countless tiny changes, one after another; each of these changes could spread among the population only if that change, on its own merit, enhanced the fitness of an animal that carried it—that is, the number of reproducing off-spring that it had. Yet obviously, the changed scale became a less efficient scale long before it turned into an efficient feather.

One day we encountered an article about the amaz-ing structure of the feather, with stunning photographs taken through an electronic scanning microscope. Could it be, asked the authors, that such a marvelous structure evolved through gradual changes? Or rather did a higher Providence create the feather for the func-tion it has today? It turned out that the article was published by a fundamentalist foundation to disprove evolution and to present creationism as a scientifically valid paradigm. Of course, present-day natural scien-  tists dismiss creationism, but we still confront the question: How could a scale evolve into a feather?

Signal selection works counter to utilitarian selection; it therefore favors a series of changes that decrease rather than increase the straightforward usefulness of a body part. The selective advantage of each of these changes is not efficiency but rather effectiveness as a signal that conveys reliable information. It would be en-tirely logical to posit that some scales changed gradually into complex featherlike decorations. A heron nowadays survives well even though some of its feathers have become threadlike decorations—signals—and no longer help the heron to fly or to insulate its body. In just the same way, some of the scales on an ancient reptile's front legs might have evolved into loose, long, decorative plates that impaired its ability to walk and climb, and were favored by signal selection precisely because such a handicap was reliable evidence of the reptile's walking and climbing ability. Such decorative scales might eventually turn out to be useful for gliding from tree to tree—opening the door to the evolution of feathers by utilitarian selection.

 This is not a new idea. Many before us have recognized the crucial role of "sexual selection" in producing novel trends in evolution—and as we have seen, what Darwin termed *sexual selection* is a subset of the process we call signal selection. Our discovery is not the process itself but rather the logic behind it. We say that by its very nature, the reliability required in signaling militates *against* efficiency. Handicaps increase the reliability of signals not *despite* the fact that they make an animal less efficient, but *because* they do. Any im-provement in a signal *must* be accompanied by a cost to the signaler—that is, it must make the signal's bearer *less* well-adapted to its environment.

Handicaps are not random. Specific handicaps evolve as signals in order to demonstrate specific qualities and abilities reliably, and this channel of natural selection—signal selection—runs contrary to simple selection for efficiency—utilitarian selection. Through this channel can come the first stages in the evolution of new body members, which start out by serving as signals, and which then acquire new functions by gradual selection for efficiency. We believe that it is through this alternating process that the deer got its antlers and the bird its feathers.

CHAPTER 8

THE USE OF COLOR FOR SHOWING OFF

Color may camouflage an animal, in which case its benefits are obvious and need no explanation. Color may also affect heat absorption or reflection. There are cases in which coloring itself signals health, as do the red of cocks' combs, the color of human lips, and similar structures that show off blood circulation, and the color of flamingoes and other birds that get their hue from foods rich in carotenoids.[1] Carotenoids are active molecules and can cause damage within the body; we speculate that high levels of carotenoids in the blood at the time the bird is growing its feathers may well be a handicap that can be sustained only by a high-quality bird.

But color, unlike markings, rarely conveys a specific message. Rather, color determines the conspicuousness and precision of messages expressed by markings, decoration, shape, and movement. The message may be bold and striking, clear even from a distance, or cryptic and subdued, readable only at close quarters.

Traveling in different habitats over several continents, we watched birds and animals and constantly asked ourselves: What is it that we see in each? What is the animal showing off? What would have changed in our perception of that particular individual under these particular circumstances if the colors of its markings were different? We tried to understand how certain colors appearing over an animal's whole body, or on large parts of it, affect the distance at which a shape or pattern can be seen, and the clarity of the message.[2]

BLACK IN THE DESERT

The first inkling we had that there might be a logical reason that a particular bird was a certain color came from a paper published in 1950 by two outstanding ornithologists, Mayr and Stresemann.[3] They pointed out that wheatears living in open habitats, such as the desert wheatear and the isabelline wheatear, are cryptic in coloring, while wheatears that live in hilly country and inhabit deep gorges, such as the mourning wheatear and the white-crowned black wheatear, have conspicuous black and white coloring. They also pointed out that the tails of species living in open habitats tend to be black, while the tails of those living in mountains have more white in them.

We soon noticed that other desert birds tend to have black body parts that in birds of more temperate habitats are usually white: skylarks have white on the sides of their tails, while the sides of desert larks' tails are very dark (though not quite black). The Cretzschmar's bunting in the Mediterranean areas of Israel has white tail feathers, while the desert bunting has no white in its tail.

This rule applies to desert mammals as well. Male ibexes have huge black horns and black beards, both very conspicuous against the beige background of the desert; sometimes the black horns and beard are the only clue to the presence of this otherwise very cryptic animal. During courtship, the male ibex grows magnificent black fur on its chest. Black shows up clearly in the desert in all lights, except for the very rare cases in which the desert background itself is very dark. Black is more clearly visible than any other color even in the middle of the day, when the hot air blurs one's vision. At dusk, when the gray body of a blackstart merges with the twilight, the spreading black fan of its tail can still be seen.

In the deserts of the Negev and the Sinai, the black garments and black goats of the Bedouin are conspicuous from far away; it occurred to us that this conspicuousness might be precisely the reason these people wore black clothes and chose black goats. When we mentioned this idea to researchers who study the physiological connection between colors and temperatures, they smiled and pointed out that in the Sahara, the Bedouin dress in light colors; wouldn't this indicate that the Bedouin select the color of their clothing arbitrarily?

When peace with Egypt allowed us to visit the Western Desert, the Egyptian part of the Sahara, we realized that there was a crucial difference between it and the desert we knew in Israel and the Sinai. The Sahara is far dryer, and some parts of it see no rain whatsoever for years on end. In the Negev and in the Sinai, Bedouin live with their herds all over the desert; in the Sahara, they live in crowded oases. They may go into the desert to travel from one oasis to the next, or to make use of temporary grazing grounds that appear after unpredictable rains; but they don't live in the open desert.

Since the ecological and social conditions in the Sahara Desert are so different from those in the deserts of the Negev and the Sinai, it follows that the background colors, the distance a message must carry, and perhaps even the actual messages conveyed are also different. In short, the fact that dwellers in these two different deserts favor different colors actually supports our claim that there is a relationship between the environment and the colors of advertisement. One needs to know the advertiser, the message conveyed, and the conditions under which it is conveyed before drawing hasty conclusions about the use of certain colors.

BLACK AND WHITE IN OPEN SPACES

Black and white are the dominant colors in open spaces: in the deserts, on steppes, at the seashore and on the open sea, and in the sky above the canopy of tropical forests. It is unusual in these spaces to see red, yellow, blue or orange birds (though there are some exceptions, like bee-eaters and rollers). Black and white are both used for long-distance advertisement; each of them has its benefits and drawbacks.

What is the difference between black and white, and what is each good for? We hit upon the answer while driving on a highway. Along the road stood telephone poles capped with small porcelain insulators. Some of the insulators were black, others white, in no particular order. We thus had a chance to look at black and white objects of the same shape and in the same light. Sometimes the background was the sky, sometimes dark rocks, sometimes greenery. We soon realized that we could see the white insulators from a greater distance, but that we could see the shape of the black ones more clearly. A white surface is seen over longer distances, since it reflects most of the light hitting it; yet that very reflectiveness seems to cause the edge of a white body to look somewhat blurred. The edge of a black object, on the other hand, appears bolder, probably because it absorbs all the light that hits it, thus creating the greatest possible contrast between the object and its background, which usually reflects some light.

Birds that advertise their presence by perching for long periods in high places tend to be black: ravens, drongoes, black eagles, and so on. Since black provides the greatest contrast with most backgrounds, it shows these birds' shapes to perfection. The blurring of the edges of white bodies may explain why hardly any small birds are completely white. Presumably, such birds as swans or egrets are large enough that a minor blurring of their outlines is insignificant for them.

White is especially effective in advertising movement: a soaring flock of pelicans or gulls is revealed by the light reflected off the white patches on their wings and bodies as they turn. The reflected light can reveal speed, wing movement, and other factors. In fact, many birds have white in their wings or tails that is visible only when they are in flight. The blurring caused by white is not an issue here, since these birds' motion itself blurs their outlines.

In Australia, we visited friends whose daughter came home crying from her first ballet lesson because the teacher scolded her for not wearing the traditional black leotard. The teacher explained that without the black leotard, the exact position and line of the girl's body could not be seen as clearly and so were harder to correct. In performance, on the other hand, ballet dancers are more likely to be attired in white or in very light colors: this increases the audience's ability to appreciate the dancers' movements rather than isolated positions.

COLORS IN FORESTS AND ON CORAL REEFS

In the tropical forest, one finds red, yellow, and blue birds; few are black or white, and it usually turns out that those few spend significant amounts of time above the canopy or on the ground, outside the patchwork of light and shadow that characterizes the canopy itself. In the dense vegetation, birds see each other only over short distances; white and black thus lose most of their advantages. Indeed, the patchwork of gleaming rays of sun and deep shade within the canopy distorts white, black, and black-and-white shapes and patterns.

But there is still a need to show off both shape and movement. Yellow and red are easily distinguished both from patches of sunlight and from the shadows of leaves and branches. Yellow has most of the reflectiveness of white, reflecting all light colors other than blue, yet it is different enough from white not to get mixed up with rays of sunlight; various shades of yellow show off movement in the forest. Red contrasts best with the green canopy. It reflects little light and, while dark, is different enough from patches of shade not to be mixed up with them; forest birds that advertise by perching are red. On the forest floor, where the light is dim and there are very few patches of sun, white and black are again the dominant colors for advertisement. We also find a great deal of green in forest birds—but though this color looks bright and prominent in zoos, in the birds' natural habitat it is actually cryptic and conceals the bird against the green background of the forest.

The influence of environment on the selection of color may explain a phenomenon noted by Diamond[4]: mixed-species flocks of birds in tropical forests are usually made up of species that have similar colors. He describes flocks of birds who are mostly black and brown on the forest floor; green and yellow birds, mostly insect-eaters, wander ten to twenty feet up. Each such flock roams a certain level of the jungle, and we think the birds' colors suit the part of the forest in which they are active: the top of the canopy, halfway down, or the forest floor. Since birds of whatever species on a given "story" have to deal with the same conditions of light and distance, they use similar colors.

We find the same pattern among fish. In open water most fish are gray, black, or white; on the coral reefs and in kelp forests, where long-range visibility is obscured and shadows and patches of light abound, bright colors enable fish to show off at short range.

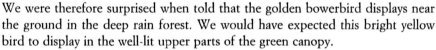

Once we figured out these "rules," we made up a game, using the descriptions of birds' colors in guidebooks to guess the habitat and some of the behavioral traits of species unknown to us. When we compared our predictions with the knowledge of local birders, we usually found a good match. We were therefore surprised when told that the golden bowerbird displays near the ground in the deep rain forest. We would have expected this bright yellow bird to display in the well-lit upper parts of the green canopy.

Fortunately, we were able to visit Gerald Borgia at his study area near Atherton, Australia. We waited for a long time in blinds near the bower of a golden bowerbird, hearing its calls without seeing it. Suddenly the bower was lit by a ray of sunlight. The male came flying down the sunbeam as though he were part and parcel of it, and landed on his perch, a horizontal branch joining the two pyramids of sticks that made up his bower. The light was reflected onto the bird by light gray lichens, which he had placed on the same branch. The bowerbird's golden color now made sense; but how could sunlight penetrate that deep into the forest floor? We later learned that male golden bowerbirds meticulously pick off the leaves above their bowers to let the light penetrate at the angle they desire.

THE USE OF TWO COLORS

Blackbirds, ravens, and cormorants are all black; egrets and swans are all white. But more often birds are two colors or more. Two contrasting colors side by side—often on two distinct components of the same body part, such as a tail or wing, as we saw in chapter 5—can show off shape and relative dimensions. Parts of the body that are seen while the bird is perching are most often black; white and light colors are often hidden when the bird is at rest and revealed only when it is in flight. Some species, though, have markings of both colors that can be seen when the bird is at rest. This allows the bird to advertise in all lights: the white belly of the great gray shrike and of the wheatear glisten in the light when the sun is at the observer's back, but when the bird is perched between the sun and the observer, the black at the side of its wings helps define its outline. Sometimes the messages are aimed only at nearby observers: the zebra's stripes, which accentuate bodily features, merge into a cryptic gray at a distance.

Two contrasting colors on a body reduce the overall impression it makes and the distance at which its shape can be seen distinctly; it may also distort the overall shape. For that reason, we often find a "compromise color": one that is dark enough to show the animal's shape, yet reflects more light than black or dark red. In open habitats and above the forest canopy the most common compromise color is gray; inside the forest it is mostly beige, orange, or brown. On the forest floor one finds a great deal of reddish-brown, which is the color of many forest mammals including squirrels, the common red dog of India, and the sambar, an Indian forest deer, all of whom live in the depths of the jungle. The sambar's undertail "flag" is light brown, unlike the white "flag" of deer in open spaces.

Another recurring pattern is the presence of two similar but not identical colors to advertise one message over a short distance and another over long distances. A nearby observer sees the details; over distance, both colors merge into one message. The distance that the detailed message travels varies with the size of the patches of color and with the intensity of the contrasts. The dark red patch on the red-winged blackbird's wing is very prominent at close range but does not lessen the effect of the bird's silhouette when it is seen from a distance.

GLOSSY COLORS AND MOVEMENT

Movement is very effectively shown off by the glossiness that occurs when incident light is reflected in a single direction instead of being widely diffused. Glossy colors are produced not by pigments but by physical properties of the feathers, scales, or cuticle. Because of physical constraints, most reflective colors in nature are blues and greens—the shortest waves in the visible spectrum. Since the reflection of light is directional, a surface that looks glossy from one direction reflects no light and therefore looks dark from all other directions. Because of this attribute of glossy surfaces, the same coloring can advertise shape and movement: small, fast-flying birds such as hummingbirds and sunbirds are glossy; the dark color defines their shape clearly, while the glossiness provides information about their movement.

EXCEPTIONS TO THE RULES

There are exceptions to the rules we have outlined: bee-eaters and rollers are colorful birds that live in open habitats; many finches and buntings (called sparrows in the New World) that live in open habitats are red and yellow, and there are black and white flycatchers that live in woodland and forests. Once one knows the bird, and what its habitat is like in all seasons, however, one can often explain such exceptions. For example, the yellow-billed finch (*Melanoderma xanthogramma*) that lives in open habitats in the Patagonian Mountains of South America has a black chin with a yellow outline, rather than a white outline, as one might expect a bird living in the open to have. Watching the finch in early spring, when the birds arrive to stake their territories, however, we realized that at that crucial time the birds often stood and sang on rocks where there were many patches of snow. White would have been less effective against such a background; yellow worked much better. A related species, *M. melanoderma,* which lives at lower altitudes, below the snow line, has a black chin outlined with white.

Another example: the black and white flycatchers (Picedula spp.) arrive at their breeding grounds when many trees of the forests and woods are still bare of leaves. A closer look at the places in which they display may well show that at the start of breeding season, when effective display is crucial, the light in their habitat is similar to that of open country—even though they are forest birds; thus the use of black and white.

Another possible complication is the effects of ultraviolet light, to which some birds are sensitive, unlike humans.[5] It is very possible that some colors look different to birds sensitive to UV than they look to us. This may explain the colors of kingfishers and bee-eaters. On the other hand, the amount of UV light is usually minimal, especially in forests. This subject requires more detailed research.[6]

The hypotheses presented in this chapter are based on our subjective impressions of the birds we watch. Still, the search for logical explanations regarding the colors of particular birds and animals has taught us to see their lives and habitats with new eyes. Our observations tend to raise questions other than those dealt with in most studies of color, and we hope our suggestions will lead to new and fruitful research.

CHAPTER 9

CHEMICAL COMMUNICATIONS

Many organisms, both mono- and multicellular, communicate by secreting chemical molecules and make decisions based on chemical molecules secreted by others. Such chemicals, which are produced by one individual and influence the actions of other individuals, are called *pheromones.* In order to carry reliable messages, pheromones have to meet the three conditions we have established for signals: they have to be costly to the individual sending them; the cost has to be more of a burden to a dishonest communicator than to an honest one; and there must be a logical relationship between the specific cost of the signal and the message conveyed by the signal.

PHEROMONES IN BUTTERFLIES AND MOTHS: CHEMICAL HANDICAPS

We touched briefly on chemical communications in chapter 3, when we discussed arcteid and danaid butterflies. Many other butterflies of the Danaidae and Arcteiidae families secrete pheromones that contain chemicals—often poisonous ones—that accumulate in their bodies from the plants they eat, or that they collect in other ways as a defense against predators.[1] Such poisonous chemicals or their derivatives in a male's pheromones testify that he is able to deal with the poisons and utilize them.

Male pheromones vary greatly from species to species, reflecting to a degree the varieties of plants they eat. Female pheromones, on the other hand, are much more similar.[2] Researchers assert that female pheromones evolved in order to enable males to locate and identify females of their species who are ready to copulate. But if that is so, it is strange that the female pheromones of different species are so similar. At the least, one would think, female pheromones would vary as male pheromones do.

In the discussion of species-specific markings in chapter 4, we explained our conviction that markings that are *common* to a group, and have therefore been thought to identify the group, actually evolved out of *competition* members of the group engage in to establish the superiority of one individual over others. We believe the same principle applies to pheromones that evolve in females of a given species. It makes sense that females compete with one another over males, each trying to attract males and mate with the better ones. Since males indeed follow pheromones to find females, the pheromone molecules presumably give reliable information on the quality of the female disseminating them. If that is the case, then there must be a true correlation between the quality of the female and the quality or quantity of the pheromone she secretes; the cost of the signal has to be high enough to prevent would-be cheaters from abusing it. In short, the pheromone has to be a chemical that would harm a female who overproduced it—and the level a female can safely produce should correlate to qualities that are of interest to males.

What kind of harm can overproduction of pheromones cause the female? We are now conducting theoretical research to answer this question. At this stage, we believe that the pheromones concentrate around the female that produces them, penetrate her sensory cells, and impair them.

Female pheromones are mixtures of molecules, mostly long, unsaturated chains consisting of ten to twenty carbon atom groups, that usually contain at their end an active chemical group that is water-soluble: an acetate, acid, alcohol, or aldehyde. It makes sense that the long chains of the pheromone, which are fat-soluble, can penetrate the fatty membranes around the female's sensory cells. The sensory receptors on these membranes are proteins with a particular spatial configuration based on a precise folding of protein chains; we speculate that the active segment of the pheromones—the acid, alcohol, or aldehyde—may damage the receptors that let females sense various chemicals around them and thus impair the female's ability to react to her surroundings. The secretion of too much pheromone would thus dull the female's senses. This dulling of the senses would impair her ability to evaluate the males courting her—and what is the point of attracting males if one cannot distinguish among them? Indeed, Ellis and his colleagues found that female cotton leafworm moths who were placed in an environment high in female pheromones stopped reacting to male courtship; the number of copulations fell almost to zero.[3] Male activity, on the other hand, was increased under similar conditions.[4]

What is the connection between the secretion of female pheromones and the quality of the female? It may well be that a female who is better able to manufacture new membranes to replace those impaired by the pheromones can therefore afford to secrete more pheromones. It may also be that the ability to replace membrane is correlated to the fat reserves in a given female's body, which in turn correlates with her ability to lay eggs. In that case, the production of more pheromones, or of pheromones with a higher concentration of active molecules, would testify reliably to the female's ability to lay eggs, making her desirable to males.

The evolution of female pheromones probably parallels that of any other signal: to begin with, females secreted various fatty acids or related fatty substances on the cuticle, like many other organisms. These chemicals were costly to a greater or a lesser degree, partly because of the damage they inflicted on living cells; but the protection they offered was worth the cost. Still, the amounts each female could afford to secrete would vary, and it is reasonable to assume that there was a correlation between the quantity of these chemical secretions and the quality of the female secreting them. Quite likely, males evolved receptors sensitive to the chemical mixtures that best reflected a female's quality. (This would be analogous to the "mind-reading" in the development of ritualization, at which stage observers note functional motions or postures that precede certain actions but which are not intended to signal those actions.) Once that happened, then mutations causing overproduction of these chemicals would benefit their bearers. At that point, signal selection took over, and picked as pheromones the chemicals that impose the most telling handicaps and thus provide the most reliable information about the quality of the female.

Although there is a great deal of similarity between the pheromones of females of various species, each species has its own special mix of chemicals, which can be used to identify it. The male's brain has receptors adapted to the specific chemicals typical of a female of his species; a few molecules of chemical are enough to cause a reaction in the male's brain. This would seem to support the idea that the pheromones evolved in order to identity the species of the female, but we think otherwise. After all, the females themselves have receptors that react to chemicals secreted by the plants they lay eggs on, plants that their offspring consume; yet it is highly unlikely that the plants secrete the chemicals in order to announce their presence and their identity to their pests!

We think that the typical mix of pheromones for each species is a by-product of the species' own distinctive metabolism. We can draw an analogy to alcoholic beverages that humans consume. Each nationality has its characteristic drink: Russians typically imbibe vodka; Serbs prefer slivovitz, and Japanese like sake. But nobody claims that the various alcoholic beverages evolved *so that the nationality of their consumers could be determined.* Rather, each developed out of a people's simple desire for an alcoholic beverage. The special character of each beverage is a by-product of the difference in the raw materials and technology available in each culture.[5]

The special physiology and diet of each species, then, affects the composition of the female pheromones of each species. But that substance evolves *as* a pheromone in the same way that other signals evolve: it delivers a message about the genuine quality of the individual that secretes it in a way that cannot be profitably faked. We assume that male moths choose their mates according to females' ability to secrete harmful—that is, handicapping—components of their pheromone. Obviously, as in other cases of selection, moths do not rely solely on the composition of pheromones to choose among candidates; other factors, such as the rate and timing of pheromone emission, and the female's movements, probably affect the final selection of one female over another.

YEAST SEX PHEROMONES AND PROPHEROMONES: THE ROLE OF GLYCOPROTEINS

Thus we see that primarily, pheromones advertise an individual's quality, rather than tagging him or her as a member of a specific group. This led us, together with Edna Nahon, Daniella Atzmony, and David Granot to develop a new interpretation of sexual pheromones in yeasts.[6] That interpretation is still largely speculative. But it deserves to be put forth here because it illustrates how the Handicap Principle may apply to communication among single-celled organisms just as it does to signals among more complex animals.

Yeasts are unicellular organisms. Yeasts of the species *Saccharomyces cerevisiae* (baker's yeast) have two genders, *alpha* and *a,* that mate with one another. Each of them secretes a special peptide, and has a special receptor for the peptide secreted by the other. Detecting the peptide of the opposite gender leads the individual receiving it to stop growing and begin the processes that prepare it for mating. Mating involves merging the two cells together to form one organism. Mating in yeasts, as in other unicellular organisms, is an all-or-nothing proposition: each has only one chance to mate in its life.

Up to now it has been assumed that the peptide simply announces the presence of an individual of one or the other gender that is ready to mate. But recently Jackson and Hartwell[7] found that yeasts select the cells they mate with, and prefer to mate with one that secretes more pheromone. They suggested that the amount of pheromone secreted by each yeast cell advertises its quality.

The fact that yeasts select their mates rather than mating at random supports our assumption that each organism advertises its quality, rather than its identity. But we doubt that the *amount* of peptide constitutes a reliable test of quality. The short peptide that comprises the pheromone is not a harmful chemical; its raw materials are readily available; and every yeast cell is genetically programmed to produce it and supplied with the necessary information. The process of protein production in the cell is an ongoing one, and the pheromone peptide constitutes only a minuscule part of the yeast cell's output.

Thus, it is hard to imagine how a short, simple peptide can reflect the phenotypic quality of the advertising cell. In addition, the concentration of a chemical is not a good indication of quality, since a nearby cell producing small quantities would be perceived as equal to a more distant cell producing much higher quantities. So we started looking for indications that other molecules are involved in the communication between yeasts, molecules that can testify to the phenotypic quality of the cell by means other than sheer quantity.

It turned out, when we surveyed existing literature, that the yeast pheromone—the peptide—is cleaved out of a large protein molecule, called a *propeptide* or *propheromone.* The propeptide of the alpha gender is a glycoprotein—a protein that carries sugars. We believe that glycoproteins are better suited to testify to the quality of a cell than simple proteins, or peptides. Unlike the manufacture of the amino acid backbone of proteins, all of which are produced by the same enzymatic machinery, the synthesis of the special sugars of glycoproteins occurs by means of special enzymatic pathways that are activated only under specific conditions, which depend on the physiological state of the cell. For example, in humans the composition of the various sugars on the gonadotrophic hormone FSH, which is a glycoprotein, varies with changes in the female menstrual cycle.[8] The composition of sugars on a given glycoprotein and the number of units of sugar vary. This microvariation, as it is termed, can reflect specific conditions in a way that pure proteins cannot. In other words, the complement of sugar units on a specific glycoprotein can reflect reliably the condition of a cell.

In yeasts, it turns out that the alpha propheromone is essential to the process of mating, not just necessary to the manufacture of the pheromone: in lab experiments, yeast cells that are manipulated so that they can produce the alpha propheromone but lose the ability to cleave the pheromone from it can still mate, if pheromone is added to the mixture; but when scientists prevent yeast cells from producing the propheromone, the cells cannot mate even if supplied with the alpha pheromone.[9]

Two genes produce alpha propheromone, and each produces a different version: four molecules of pheromone are cleaved from one version of the alpha propheromone; an identical pheromone molecule and one that is slightly different are cleaved from the other version. All natural populations of yeasts have both genes, even though the pheromone cleaved out of the two versions of the propheromone is nearly the same. It would seem that the order and number of molecules of pheromone in each of the two variants of the alpha propheromone are essential and important, not accidental.

What is the role of the alpha propheromones in the communication that precedes mating between yeast cells? One end of the propheromone has a hydrophobic segment—a segment that dissolves in fats, but not in water. The other end contains the pheromone segments. The sugars of the glycoprotein are connected to a middle segment. The pheromone segments are cleaved off the propheromone by enzymes and are secreted by the cell. The pheromone molecules reach the other gender, are sensed by its receptors, and trigger the receiving cell into sending a copulating offshoot toward the cell secreting the pheromone.

There is indirect evidence that complete molecules of the propheromone—that is, propheromone with its pheromone segments still attached—can be found on the outer membrane of the yeast cell that secretes the pheromone. Our conjecture is that the alpha propheromone is anchored to the cell membrane by its hydrophobic segment in such a way that the pheromone sections attached to it protrude outside the cell membrane, and that they are held in a specific spatial configuration—in which the sugars in the middle segment of the molecule probably play a role.

We think that these pheromone segments may bind to receptors in the other cell's offshoot, just as free pheromone does. It may even be that both types of the alpha propheromone are needed for a good link: a bond forms between the two alpha propheromones (a heterodimer), and then their peptides link with a group of receptors on the membrane of the receiving cell. This would explain why the pheromone is manufactured by two genes rather than one. It makes sense that impairments in the genetic or phenotypic quality of the advertising cell would affect the number of sugar molecules attached to the propheromone and thus the spatial configuration of the propheromone in that particular yeast cell's membrane. A change in the spatial configuration could well affect the strength of the bond between the propheromones on the signaling yeast cell and the receptors on the other yeast cell, and thus affect mate preference: an individual cell weakly bonded to its would-be partner might detach itself to bond strongly to a preferred mate.

If this conjecture proves true, then the role of the pheromone peptide itself—a relatively small, mobile molecule—would be merely to draw the attention of cells of the other gender to the cell that secretes it, so that they would come and "investigate" by sending a mating offshoot toward its membrane. The critical "decision" whether to go ahead and mate with that particular partner would be based not on the pheromone but on contact between the propheromone on the membrane of the one yeast cell and the receptors on the other cell's membrane.

We have described our interpretation of the communication between yeast genders in detail because we see it as a general model: it illustrates how cells can advertise their phenotypic quality by means of peptides and their propeptide progenitors, which hold them in specific spatial configurations. Indeed, recent findings show that the propeptides of other peptides used in communication also bond to the same receptors to which the peptides themselves attach.[10]

We believe that most if not all chemical communication between individuals contains reliable testimony about the phenotypic quality of the one sending the signal. Molecular structures can serve this purpose no less than bodily adaptations do in highly developed organisms such as mammals and birds. Needless to say, we do not claim that yeasts make conscious decisions; but the process of natural selection saw to it that individuals who followed better life strategies, by means of whatever mechanisms, survived and multiplied, while those who did not perished.

CHEMICAL COMMUNICATION WITHIN THE MULTICELLULAR BODY

Unlike unicellular organisms, all the cells in a multicellular body have the same interest. It would seem at first glance, therefore, that within such multicellular organisms there is no need to test the reliability of communication between cells, that such communication need only be clear and efficient. Yet it turns out that the molecules used in communication between cells within the body—hormones, growth factors, and neurotransmitters—are similar to, and sometimes identical to, the molecules of pheromones used in chemical communication between separate organisms. Many of these hormones are glycoproteins or harmful chemicals—the same kinds of chemicals that, in our opinion, serve as the handicaps that guarantee that chemical communication between individuals is reliable. The question is, what is the role of such handicaps in communication within the multicellular body?[11]

The multicellular body has a division of labor: the special role of some cells is to receive information from their surroundings, process it, and pass on instructions that change the behavior of other cells in the body. These instructions are sent by chemical signals—hormones. If for any reason these instructions are wrong, the resulting harm will affect all body cells, including both the sender and the receiver of the signal.

The genes in all the cells of a given individual's body are basically identical, of course, and so the genes for manufacturing all chemical signals used in the body exist in all body cells. The difference between the cell whose role is to send the signal and other cells is not one of genotype—all cells in the body are of the same genotype—but rather of phenotype: where the particular cell happens to be in the body, and what its state is. A mistake in sensing, in decision making, or transmission could cause a cell to manufacture the wrong chemical signal or to send that signal off at the wrong time. Thus, in order to prevent mistakes, the signal itself should testify in a reliable way to the phenotypic state of the cell sending it. A signal within the body can be reliable only if it cannot be produced by cells of the wrong phenotype. In other words, we find a use for the Handicap Principle within the multicellular body.

It was recently found that free radicals like nitric oxide and carbon monoxide, which are very poisonous, are used in communication between cells in the body and are secreted by nerve cells in important body parts like the brain and the heart.[12] This use of highly toxic chemicals in the most crucial bodily messages supports our view. On the other hand, the fact that many hormones and many neuropeptides that convey signals within the brain are short peptides seems to contradict the Handicap Principle: as we said in the discussion of yeasts, we don't think that short peptides have the capacity to provide reliability in communication.

It may be, though, that in at least some of the peptide signals in the body, reliability is provided not by the short peptide itself but rather by the prohormone (propeptide) to which it is attached.[13] The propeptide, in other words, may well provide reliability in the same way that we suggest the propheromone does in communication between yeast cells.

Another method for ensuring reliability in signaling is used, it seems, in the EGF system. EGF is a peptide that enhances the division of cells in the walls of blood vessels. These cells divide when a damaged blood vessel needs repair. It turns out that EGF causes cells to divide only when it is attached to special sugars. EGF and the sugars attach to one another in the intercellular substance in the walls of blood vessels; when the tissues are injured, EGF and the sugars are released together. The presence in the blood of EGF attached to sugars is a reliable indicator that blood vessels have indeed been injured and that there really is a need for repair.

Like any signals, chemical signals between competing organisms have to be reliable—otherwise they will not be accepted by potential receivers. Chemical signals within the multicellular body also demand reliability: not because of conflicting interests, but rather to prevent mistakes.[14] If we are right, then all chemical signals—including signals within the body—work according to the Handicap Principle. It should therefore be possible to find a logical relationship between the structure of molecules used in signaling and the messages they carry. This premise guides our current research into the messages encoded in molecules that are used in chemical communication.

PART III

THE HANDICAP PRINCIPLE IN SOCIAL SYSTEMS

TESTING THE BOND

At the end of each workday, when we arrived at home, our little daughters used to come running to welcome us. They would jump up on us and demand that we take part in their games or at least tell them a story, no matter how exhausted we were. Our dog, too, would leap up and insistently push himself against us. And in the mornings the dog would make even more of a nuisance of himself: he would lie down in the doorway, forcing us either to walk around him or to push him aside every time we went from the kitchen to the dining room.

All of us, including the dog, knew that we belonged together and were willing to do a lot for each other. But even in the most loving family, the *degree* of willingness and ability to cooperate may well change from day to day. Thus, one of the most important questions for each member of a partnership is the degree to which other partners are willing to do things for it at any given time. The eternal questions "Do you still love me?" and "How much do you love me?" are especially urgent when one of the partners has been away for a time, or is getting ready to leave. Our children and our dog revealed to us the existence of behavioral systems that enable individuals to collect information about the social bonds between them and their partners.[1]

TESTING BY IMPOSITION

How can one get reliable information about the quality and strength of a social bond? Actions that benefit another are likely to be accepted by him or her for their own sake. The only way to obtain reliable information about another's commitment is to impose on that other—to behave in ways that are detrimental to him or her. We are all willing to accept another's behavior if we benefit from it, but only one truly interested in the partnership is willing to accept an imposition. As we shall see, all mechanisms used to test the social bond involve imposing on partners.

People who don't like dogs will not let a dog jump up on them or lick their hands. Almost anyone would find it too much of an imposition to be jumped on by a large dog like a pointer or a German shepherd. Such dogs use their weight to test the bond: they approach and stand next to a visitor in a friendly manner, leaning against the visitor's legs, gradually transferring more and more of their weight, until they are pushed off. How much of this treatment a visitor will take before pushing the dog off enables the animal to assess the visitor's attitude toward it. When a dog gets in its owners' way in the morning, the way they move it aside or go around it when they are in a hurry gives it the information it seeks. And at the end of the workday, a dog can find out reliably within a minute how much attention its owners are willing to give it just then, by how willing or unwilling they are to be jumped up on.

Information about the importance of a partnership to each of its members is useful on several levels. In the most extreme cases, it may lead to a decision to break up the partnership. Most of the time, though, it helps individuals decide how much to invest in a partnership and how much can reasonably be expected from the other partner. Clearly, there is no point in asking for something from a partner who has no interest in the partnership.

But even among willing partners there are often conflicts of interest. A cake eaten by one is not available to another. A moment of attention given to one partner is a moment not devoted to a third. And even when all members of the partnership benefit equally from an action, if one partner carries out the action when another might have, it can be seen as a loss to the one and a gain to the other.

The importance of a partnership to its members and the relationships among partners depend on many factors that can easily change, which makes frequent testing of the social bond essential. It is important to each to be aware of other partners' attitudes at any moment: it can be costly to make a request at the wrong time, because once it has been denied, it may be less likely it will ever be granted. It is better to assess things first and wait until the odds are on one's side.

• • •

AGGRESSION IN COURTSHIP

An individual can test a social bond only by imposing upon another. The degree of imposition depends on the importance of the information sought and on the state of the bond. During early courtship, the risk involved in a decision to mate is very high, and the preexisting investment in the partnership small. Under such conditions, the impositions tend to be high.

Some courting males behave aggressively toward females; most females respond by leaving, and the male thus finds out which ones are truly interested in him. Such aggression is especially common among birds when both the male and female invest a lot in their offspring. In many such species, males first establish territories and then wait for females to join them. The male attacks every female who shows up in his territory, and the females fly away. Some of the females come back again and again. Gradually, the male's aggression toward one of the females lessens and then disappears completely or almost completely for the duration of their lives together. Israelis are familiar with pairs of wagtails, small European songbirds that winter in Israel; the wagtails go through this process every fall, though they do not breed together; they only cooperate in defending their feeding territory,[2] and each of them travels north alone in spring to breed in Europe with a new partner.

As we see it, this high level of aggression early on is the simplest way for the male to test the intentions of females who are proposing to share his property (his territory) and his future. A female might find it convenient to settle in a territory on a temporary basis, so as to have a  secure place to eat and reside while searching for a permanent mate among the neighboring males. Such a temporary arrangement would benefit the female but could harm the male considerably, because during that time other desirable females who might otherwise choose him would not approach him. By chasing a newly arriving female, the male tests her intentions. A female who regards him as a temporary convenience will not put up with being chased as much as a female who has already decided that this territory and this male are her permanent choice.

In textbooks this aggression of males toward females is explained as resulting from the fact that males have to be highly aggressive in order to chase rival males away from their territory;[3] they cannot risk accepting females, whose appearance is suspiciously similar to that of the rival males. In time they learn to accept a particular female and stop attacking her. This is no more than a description of the facts presented as an "explanation," and is an insult to the birds' perceptive ability. Male babblers, too, chase females who try to join their group. They continue to attack females to a lesser extent throughout their long lives together; but they hardly ever attack adult male members of their own group.[4] There is no reason to suppose that a male is unable to differentiate between males and females and limit its aggression to males only. In fact, in nonmonogamous species, where male-on-male aggression is especially high, there is very little aggression toward females.[5]

HIDE-AND-SEEK: GENTLER TESTING IN COURTSHIP

Among some birds the only collaboration between the genders is copulation; the males of such species hardly ever exhibit aggression toward visiting females. The males mate with many females and thus don't have to be very choosy about their partners—yet even here, some selectivity is called for. Not every female who comes to such a male's dancing arena decides to mate with him; some come just to watch him closely and compare him with others. Males, especially those who are much sought after by females, have to be careful not to spend too much time with any female who is not interested in mating; using aggression may frighten away females who might otherwise be willing to copulate, though, so the male has to do his testing peaceably.

Peacocks' courtship includes such gentle testing: when a peahen approaches a peacock, he turns his back to her and shows her the undecorated back of his tail. In order to mate, the female has to go around in front of the male. We think that this is a test by which the male determines the extent of the female's interest. A female who does not bother to go around in front of the male is not interested in examining his fan of feathers closely—and is probably not interested in mating either.

Bowers built by bowerbirds are known to be a means of male sexual display.[6] The size and decoration of the bower clearly testify to the male's ability. But what was it that led to the development of a certain structure for the bower? After talking with Gerald Borgia and viewing videorecordings he'd made,[7] we came to realize that the bower's specific structure may well have developed largely because it enabled the male to determine whether a female actually intends to copulate with him or not. When a female arrives at the bower, the male hides behind the bower or in it and watches the female to see whether she attempts to follow him. Throughout the courtship, the male repeatedly peeks out and then hides again. An interested female follows him as in a dance, revealing her intentions to the male. This can explain how the bower evolved out of a bare dancing arena: even the first stick placed on the dancing floor could make it possible for a male to hide and thus more reliably test the interest of females. Of course, once the structure evolved, then the size of the columns, their number, order, and decoration could serve an additional purpose—to enable the male to show off his quality.

The males of many bird species clean and maintain a dancing arena. A clean arena is evidence that its owner has the leisure to spend time on the arena and tend it. There is one species of bowerbird—the Australian catbird—that does not build a bower but instead cleans a dancing arena. Unlike other bowerbirds, the male of this species displays considerable aggression toward visiting females. It may be that since it cannot use the bower to test the female's intentions and to compete with other males, this bowerbird uses aggressive behavior to show off its strength and to test females' intentions.

Face-hiding games are a common way of testing mutual interest between a human guest and a small child. Visitors who don't know how interested young children are in their company often hide their faces behind something to see a child's reaction. A child who is interested in a visitor will move in order to see the visitor's face again, so as to determine how interested  the *visitor* is. Both children and adults love this game of peek-a-boo, without as a rule being aware that they are collecting information.

CLUMPING, AND PREENING OTHERS (ALLOPREENING)

Another kind of testing of the social bond is seen among babblers. It is crucially important to babblers to be aware of their partners' commitment to them at all times because of the dangers involved in fights between groups. Most border clashes between groups of babblers don't go beyond the exchange of threat calls or perhaps a few chases, but not infrequently they become real fights in which participants are wounded and sometimes even killed.

Such group fights are far more dangerous than the one-on-one conflicts of solitary birds. In one-on-one conflict it is very difficult for one bird to kill another. The victor is usually satisfied when the loser flees. Group fights are different. Once a babbler grapples with another, clutching it and being clutched by it, other group members may come to the aid of their partner—and a group of babblers ganging up on a single bird can easily kill it with minimal risk to themselves. A babbler charging an enemy group risks this fate, and its life depends on the willingness of others of its group to come to its aid.

Confrontations between neighboring groups are frequent, sometimes occurring several times a day. One never knows when a confrontation will end with an exchange of threat calls and when a rival group will decide to fight in earnest. Thus a babbler has to be ready to fight at all times. Its groupmates' willingness to fight for it varies; it can be affected by the arrival of new babblers in the region, internal conflicts, and so on. Therefore, it is important for each babbler to test its groupmates' commitment to it often. A great deal of the babblers' daily behavior serves this purpose: clumping, preening others (allopreening), and group dancing all evolved to enable babblers to test the strength of the social bond among group members.

Babblers often clump together, but this is not the norm among birds: individuals of most bird species tend to keep a certain distance from others. Swallows or starlings perched on a power line are always a set distance apart, like musical notes on a staff. If a lower-status individual gets too near, they threaten it; if a higher-status one approaches, they move away. This insistence on individual space makes sense: birds need a certain amount of room to spread their wings and fly off; one who does not keep its distance from others cannot take immediate flight to avoid a predator or catch prey that may appear. This freedom is given up by birds who clump, as do babblers and other species that depend on strong social bonds.

Many group-living birds clump. Bulbuls, parrots, and bullfinches, all of whom have a strong pair bond, clump in pairs. We think that clumping reflects reliably birds' readiness to invest in the partnership, because it hampers freedom of movement. It is no coincidence that girlfriends press against their boyfriends just when the boyfriend is talking to another girl: that's just the time to pressure him to prove his interest in the relationship.

Another method babblers use to test their social bond is allopreening, or preening one another's feathers. Babblers often clump together and preen each other. The babbler being preened puts itself at the preener's beak, feathers fluffed and eyes often closed to let the other preen its facial feathers fully. The recipient of the preening can hardly see in that state and obviously is not prepared to fly or flee. The preener tests the other's will to cooperate with it, and the one being preened shows off that it trusts the other to warn it in case of danger. Both, by clumping and allopreening, take on a greater risk than birds resting alone.

It used to be assumed that the function of allopreening is to preen feathers a bird cannot preen on its own, such as head feathers. But this does not explain why it is that only social birds preen each other, and why they preen also parts of the body that each bird can easily reach on its own. Preening each other's heads is simply a useful bonus of a behavior that serves primarily to test the social bond.

GROUP DANCES AND SIMILAR RITUALS

The most impressive social activity of babbler groups is the morning dance, which was researched in Hatzeva by Roni Osztreiher.[8] We believe this ritual is another means by which babblers test the social bond among group members. The morning dance takes place almost exclusively at the first light of day. It happens only once every several days, and it is not yet clear what makes the babblers decide to dance on a particular day. One of the babblers suddenly stops at a suitable "dance floor" and starts preening itself nervously or sprawling with its throat touching the ground. Sometimes there is no response, and the babbler gives up and goes off to look for food.

At other times, though, another babbler comes and clumps with the first, preening its own body. Two babblers clumping and preening are an almost irresistible invitation, and in most cases the rest of the group soon joins in. The dance can last up to half an hour and sometimes even more. During the dance, babblers press against each other, squeezing under and over and between their partners. They dance in the open, near a bush, even though they could dance more safely under trees.

The dance almost always takes place before sunrise, when the danger from raptors is greatest, since they can exploit the low light and surprise the babblers with relative ease. This is also the best time for babblers to feed, since many night creatures are still active—including termites, a favorite babbler food. In short, the dance does not happen at the easiest time and place but, on the contrary, at an inconvenient time and place, using inconvenient movements. An individual who is willing to undertake all this with its partners shows reliably its commitment to the group. The dancing of babblers reminded us of stories our parents' generation told about the dancing of members of pioneering kibbutzim in Israel in the earliest, most difficult and exhausting years: "For months we had nothing to eat, but we danced all night. . . ."[9]

Group rituals that test the social bond are not unique to babblers. Other social animals hold them too, often before other group activities. African hunting dogs jump on each other before embarking on a hunt.[10] The hunt is risky to the hunter: when it grabs the leg or neck of large prey and hangs on, its safety depends on its partners' cooperation. African dwarf mongooses hold similar rituals.[11] The need to test the bond may even explain why dogs jump all over and around their owners when setting out on their daily walk. And there are the daily greetings between humans. Kisses, handshakes, and the exchange of nods and verbal greetings all help us assess the attitude of our partners, friends, associates, and coworkers during the opening moments of each encounter. In chapter 18 we shall see how the sexual act became a means of testing the bond between partners for various animals.

CHAPTER 1 1

PARENTS AND OFFSPRING

Aparent and a child are the simplest reproductive coalition there is. At first
glance, it seems obvious that both parent and child have the same interest:
the good of the child. But if that is so, why are there so many conflicts
between parents and children?

According to Trivers,[1] the picture is less simple than it first appears. The child's
interest is to take care of itself, while the parent has to take care of other offspring
as well. The child aims to get from its parent whatever it *can* get, even if the parent
shortchanges and perhaps even loses its other offspring. The parent wants the
child to succeed, but only if its investment in that one child does not reduce the
total number of successful, reproducing offspring it can have.

A good example of a mother-child conflict is the nursing and weaning of baby
mammals. Trivers explains the cries of a baby who wants to nurse as a psychological weapon aimed at making the mother nurse it against her will. He asserts that
the baby does all it can to force the mother to nurse it, while the mother tries to
avoid nursing when nursing is against her own best interests.

Everybody knows that a screaming baby can force a mother to actions she
would not otherwise take. But why exactly are mothers moved by a baby's screams?
Labeling the screams a "psychological weapon" does not explain what it is that

makes them effective. Trivers did not take into account that behavioral mechanisms themselves evolve by natural selection.[2] If mothers who did not respond to screaming, or who did not respond as much, had had more successful offspring than ones who responded, then the tendency not to respond would have spread in the population, and the "psychological weapon" would have lost its potency. Behavioral mechanisms are only proximate factors in the evolution of traits; the ultimate factor in evolution is successful reproduction.

THREATS OF SELF-INJURY: THE WEAPON OF THE WEAKER PARTNER

Babbler fledglings, like human babies, cry loudly when hungry. An observer senses intuitively that they cry in order to be fed, and it was a long time before we realized what was wrong with this explanation. After all, the cries pose a real danger. They can be heard over a great distance—hundreds of yards. They reveal to predators the fledglings' location—and at that age the latter can barely fly, or even hop. In fact, these loud cries often help us find fledglings' hiding places.

The common explanation is that the cries tell the parents where the fledglings are and that they are hungry. But the cries are often loudest when the parents are nearby and know just where to find the fledglings. These striking facts tell us whom the calls are actually directed at, and why they force the parents to attend to the young: we believe that the cries are actually meant to be heard by predators. The fledglings say, as it were, "Cat, cat, come and get me! I am here and I don't care who knows it until my parents feed me." Once they are fed, the fledglings stop calling. The fledglings are forcing their parents to feed them by endangering themselves.

The weaker member of a partnership can blackmail the stronger by threatening to bring injury on itself.[3] Of course, this threat is effective only if the stronger partner has an interest in the well-being of the weaker party. This is exactly the situation with mother and young. By investing more in an offspring who cries, the mother risks reducing slightly the total number of her future offspring, but by silencing the complainer she improves the chances that an offspring in which she has already invested a great deal will survive, and she reduces the risk of losing it and its nestmates to a predator at one fell swoop. The mother is moved by her baby's screams because the screams constitute a real threat to her own interests.

Most of us have heard an "abandoned" kitten who has

climbed to the top of a tree and cries all night long. Kindly neighbors try to help it, assuming that it has lost its mother. We think the kitten is trying to blackmail its mother into continuing to nurse it, to prevent the kitten from wandering off to dangerous places. The danger that the kitten will expose itself to predators is greater than the potential benefit to the mother of weaning it early. And as long as the benefit of additional nourishment is greater than the risk of predation, it is in the offspring's interest to cry.

Yet the risk involved in the crying is, and has to be, real—otherwise the mother would not submit to the blackmail. So from time to time, a fledgling or a kitten may be eaten. The young of hares and gazelles, who have many predators, cannot afford to use crying as a weapon; the risk is too great. They lie quietly in a hiding place, and the mother comes from time to time to nurse them. Obviously, the mother does not need to hear her babies crying to find them. If the young of hares and gazelles have any way of blackmailing their mothers, we have not found it yet.

The risk of predation is not the only factor that influences parents. Even humans, who live far from predators, give in to their children's cries. Of course, the lack of predators and enemies "allows" babies to cry as much as they care to with no risk, which according to the Handicap Principle should greatly reduce the value of crying as a signal. So why do parents respond? Protracted crying by a baby harms the adult's prestige by making a bad impression on his or her partner, neighbors, and friends. Prestige has real value, as we shall see in a later chapter. It may well be in the parent's interest to give in rather than lose face.

OTHER METHODS OF BLACKMAIL

Crying is only one of the ways offspring extort help from parents. Sometimes young are more brightly colored than their parents: baboon babies are dark, while their parent's lighter colors merge much better with their habitat. Wild piglets are decorated with prominent stripes, while their parents are grayish-brown, like their environment. Fledgling babblers are more striking than their parents: their beaks are black, a prominent color in the desert, with a yellow base, and their breasts are decorated with brown. These colors, together with such attention-drawing behaviors as waving their wings, make it easy to find the young babblers. Like their cries, this behavior can be explained by the fact that the danger to which the fledgling exposes itself forces its parents and caretakers to devote more attention to it.

Babbler eggs are a bright emerald green, which stands out clearly against the gray-brown twigs that make up the nest. This may also be extortion: the mother may be "using" the color of the egg to extort her group partners' help in incubating them. Brightly-colored eggs are more likely to be taken by nest-raiding predators if not continuously incubated—and precisely for that reason are more likely to be continuously incubated by other group members; this benefits both mother and young.

Great crested grebes spend all their time on the water; ducks spend part of their time on land. Yet in a seeming paradox, the fledglings of ducks have thick plumage that lets them spend long periods of time in cold water, while the fledglings of crested grebes have thin plumage.[4] Grebes carry their offspring on their backs. It may be precisely because baby grebes' plumage is thin that grebe parents don't leave their offspring alone in the water even for a short period of time. Because they have thin plumage, the fledglings cannot survive in the water

for long, but the benefit may be the care they get. Similarly, the bare belly of the baby chimpanzee encourages the mother and its other caregivers to keep it next to their bodies. The same principle may account for the total helplessness of human babies.

Feldman and Eshel assert that it is difficult to see how the mechanism of blackmailing parents could evolve, since offspring who carry this tendency will suffer from it later—the tendency will be passed on to their own offspring, who will blackmail them in turn.[5] But this is a difficulty only if one assumes that both the degree of extortion and the response of parents are fixed. We think that the ability to extort help from parents is a benefit when used "wisely," in a way that is appropriate to existing circumstances. A behavioral mechanism such as we describe makes sense only if it is sensitive to outside conditions—if the young endanger themselves only when the extra care they receive is worth the risk they take. And indeed, when babbler fledglings hear their parents' alarm call announcing a predator nearby, they freeze and do not utter a sound.

And when an offspring reliably shows its need for care, there is a direct benefit to the parent; otherwise, the parent might expend unnecessary effort providing more food than it can eat, for example. Of course, the parent would prefer a demand mechanism that would not endanger the offspring—that would leave more leeway for the parent to decide how to spend its resources. But there is no clear line of demarcation between an offspring's need to reliably inform parents how much care it requires, and its ability to use the same mechanism to extort too much care.[6] It makes sense for parents to give in when they know that offspring really do need help—which is proven by the risk the young take in demanding that help—and it even makes sense for them to give the young somewhat more care than is necessary, in order to minimize the risk. Natural selection keeps both the extortion and the parental response in balance. The benefit to the offspring of the extra care, the increased likelihood that it will survive and reproduce, is a part of this balance; so is the effect of the parent's extra investment in the demanding offspring on its potential to produce other successful offspring.

EXPLOITATION OF OFFSPRING BY PARENTS

It is no less interesting to see cases in which parents use their power to exploit their offspring for their own gain. Gain in evolutionary terms means an increase in the number of reproducing offspring: all that is really meant by *evolution* and *natural selection* is that some individuals have more offspring who themselves manage to reproduce successfully, and that therefore the traits of these parents are the ones that will prevail in future generations. Thus, in evolutionary terms, exploiting one's children means using some of one's offspring to increase the chances that others of them will mature and reproduce successfully.

Among animals who take care of their young, the offspring's quality depends on their parents' efforts to guard, warm, and feed them. Parents often prefer one sibling over others. Many birds—such as owls, herons, and parrots—start incubating when they lay the first egg. This egg hatches first, and the nestling that emerges from it will naturally be bigger and stronger than its younger siblings. If food is scarce, the first nestling hatched can snatch food away from its siblings. This com-  petition over food may lead to the death of one or more of the younger nestmates, but the parents do not interfere.[7] Under some conditions, it is better for the parents to end up with a smaller number of stronger offspring than a larger number of weaker offspring who will be unable to compete against the offspring of others of their species. Similarly, Diamond[8] suggests in his book *The Rise and Fall of the Third Chimpanzee* that menopause came about to enable women to invest more care in their existing offspring, who still need their care to survive, rather than put their effort into producing and rearing additional offspring, whose chances for survival after their mother's death would be very small.

The rules of inheritance in some human societies, past and present, work in the same way: the bulk of the family fortune is passed on to one offspring, to ensure that at least that one will succeed. Splitting the legacy equally between all offspring would mean that each would receive only a small portion, and all might fail as a consequence. Usually the preferred heir is the firstborn son, who is in a stronger position to begin with, but sometimes it is the youngest who gets the largest share. In the latter case it may be that the parents have a better chance of raising children and grandchildren if the firstborn son leaves the home and seeks his fortune elsewhere, without competing for resources with parents who can still bring up more children.

Among macaques, a kind of monkey, it is the youngest daughter who inherits her mother's social position; older daughters are subservient to the youngest one.[9] The youngest daughter's preferred status comes not from her own ability but from the fact that the mother, and sometimes other higher-ranking females, inter-vene in conflicts among daughters in favor of the youngest. This preference for the youngest may lessen somewhat the chances of older daughters, but it also prevents older daughters from competing with their mother or younger sister, thus increasing the mother's overall reproductive success.

Such exploitation is practiced to the greatest degree among the social insects; we will say more about that in chapter 13. But even in less extreme cases, any child–parent situation combines conflict with common interest. The child is interested in getting the best start in life; the parent is interested in using his or her resources in the most effective way, not necessarily to produce the greatest number of offspring, but rather to maximize the number that will themselves successfully reproduce.

Each offspring tries to persuade its parents to favor it over its siblings; each parent tries to ensure that its resources are utilized in the best way, whether by devoting the most effort to the offspring likeliest to succeed, by helping the one to whom that help will make the greatest difference, or even by exploiting one offspring to benefit its sibling. In the last case, of course, each sibling will try to ensure that it is the beneficiary of such treatment, rather than the victim.

The specific details of each case depend on a species' life strategy and on circumstances of each particular situation. In most cases, there is room for communication between the parties: parents gauge the needs and the vigor of each offspring, and offspring do their best to convince their parents of the same, so as to get all the help they can. Indeed, child–parent relationships provide fertile ground for communication, and reliability is no less essential here than in any other communication among individuals. As we have said, the only means we know of to guarantee that reliability are handicaps.

BABBLERS, COMPETITION FOR PRESTIGE, AND THE EVOLUTION OF ALTRUISM

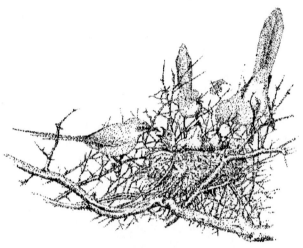

We have been studying babblers since 1970 and continue to do so as of 1997. Babblers are the only birds in Israel that we know of that live permanently in groups. As a rule, most groups have three to twelve members, though some groups have reached twenty. Babblers stay in their territories year-round and cooperate to protect them against neighboring groups of babblers and nonterritorial individuals. In breeding season all members of the group—breeding adults, nonbreeding adults, and young—take care of a single nest in the territory.

The birds' dealings with each other include a great deal of assistance to others. Such altruism is common among group-living organisms and poses a major problem in the study of evolution: it seems to run counter to Darwinian logic, according to which the individual pursues its own interest and that of its own offspring. Our study of the babblers, combined with our understanding of signal selection and of the Handicap Principle, led us to develop a new and different solution to this evolutionary enigma.

Babblers are songbirds of the many-membered tropical family *Timaliidae*. The babblers in Israel belong to one branch of this family, which spread from India northwest across the desert belt and into Morocco. The species we study is found in the Arabian Peninsula and the Sinai Desert. In Israel it can be found in the Dead Sea region, the Arava Valley, and the warm wadis (seasonal watercourses that are dry except for a few rainy days a year) in the bordering hills.

Babblers are the size of blackbirds or grackles, about 2 to 3 ounces (70–85 grams). Their tails, at about 6 inches (150–160 millimeters), are slightly longer than their bodies. The birds are grayish-brown, like their desert environment. Their wings are short; they are slow flyers, but their strong legs and long tail enable them to maneuver in thickets with great dexterity. In the open, such raptors as hawks, peregrines, and harriers can catch them easily; the babblers prefer to stay near trees and shrubs into which they can escape. They usually seek their food in and near bushes, on and under the ground, and under tree bark. They eat any creature they can swallow or tear to pieces: insects, snails, scorpions, and small snakes and lizards. We have observed two cases in which babblers killed small birds, but they could not eat them: their beaks are dull, adapted to digging in the soil rather than tearing the flesh of birds. Babblers also eat the juicy berries of some desert shrubs and occasionally eat flowers and drink flower nectar.

We study the babblers in the Shezaf Nature Reserve near the Hatzeva Field Study Center, which is run by the Society for the Protection of Nature in Israel (SPNI). We follow some 30 groups over an area of 15 square miles or so. Each individual babbler in the study area—there are about 250 of them at the time of this writing—is tagged with a combination of three colored bands and one numbered one, and can thus be easily identified. Babblers readily become habituated to the presence of humans who behave calmly and give them small bits of food; such gifts are especially welcome during winter, the lean season. They also remember well any harm done to them, however, and they stay away from whoever perpetrated it.

The babblers in the study area are used to our presence. The tiny amounts of breadcrumbs we give them occasionally do not change their food supply in any meaningful way, but as a result of these "donations" they do not fear us and let us walk among them freely and watch all their doings at close range. We can

hear the soft, widely varied calls they use to communicate, calls that don't carry more than a few yards. Some individuals are talkative, others taciturn, and still others grumble all day long. When a higher-ranking individual approaches a lower-ranking one, the latter often makes a soft sound to acknowledge the other, as if saying "Yes, sir." Adult babblers who stand guard do so quietly, but youngsters trying their hand at it vocalize softly, as if calling their nearby comrades' attention to the effort they are making.

In the sparse desert country around Hatzeva, we move easily among the babblers; we see what they are looking at and become aware of their intentions. For example, a babbler who finds food may not swallow it right away but instead may hold it in its beak and look around to see whom it can feed. If it sees only individuals whose rank is higher than its own, it may waver briefly, then swallow the food. When the beta male, the number-two male of a group, goes up to the top of a tree to stand as sentinel, we often see the alpha male, his superior, busily looking for food; he then gives it to the sentinel in full view of the other members of the group and replaces him on guard duty. In many cases, we can tell when the sentinel notices the alpha male's preparations to feed him from the direction of the sentinel's gaze, and often because the sentinel abandons his lofty post before the alpha male arrives to displace him.

Our team checks on each group of babblers in the research area about once a week, but often more frequently. Thus we closely follow changes in the composition and routine activities of each group. Over the years we have learned their life histories and social structure. Detailed studies have focused on specific aspects of babbler behavior: sentinel activity (Tirza Zahavi, Tony Larkman, Nir Faran), territoriality (Arnon Lotem), the feeding of young (Thamsi Carlisle), allofeeding among adults (Amir Kalishov), playing (the late Orit Pozis), mobbing (Zahava Carmeli, Avner Anava), allopreening (the late Andres Gutman), the morning dance and feeding at the nest (Roni Osztreiher), shouts (Zohar Katsir), water balance (Avner Anava), and competition over mating (Yoel Perl).[1] Kim Lundy and Patricia Parker carried out DNA studies, Dietmar Todt and his students are studying vocalizations, and Jonathan Wright is studying feeding at the nest.[2]

TERRITORIES, GROUPS, AND NONTERRITORIAL INDIVIDUALS

Each babbler group has its own territory. A group protects its territory against neighboring groups in daily or almost daily border encounters, which occasionally develop into clashes and at times escalate into all-out raids on neighboring territories. Groups also protect their territory against nonterritorials—babblers who have been thrown out of a group or whose group has lost its territory.

As a rule, any area with enough food and shelter to support babblers is occupied by a group of the birds. Babblers who are not members of a territorial group have a hard time finding food and shelter, let alone breeding, because the groups that protect each territory attack them on sight. The greatest danger to such nonterritorial babblers is not predators—whom they can avoid by diving into a bush—but the territory-holding babblers, who are equally adept at maneuvering in dense vegetation, and whom they can escape only by fleeing into the shelterless open desert.

Nonterritorials are never a large part of the population; we estimate them at about five percent or less at any given time. They face a high risk of predation, and most do not survive long. Still, the life of a nonterritorial, with all its hardships and dangers, is not totally hopeless. Occasionally, one of them will live for months or even years near its old territory, hiding from the territory holders, until it finds a group that accepts it as a member or manages to find a vacant area and joins with other nonterritorials to form a new group.

Babblers are long-lived and may reach the age of twelve or fourteen. In years of drought, when food is scarce, babblers do not breed. In rainy years, when the Arava Desert gets a few centimeters of rain at the right intervals, a group of babblers may produce two or three successive broods. Because of the combination of breeding success and long life, many individuals who cannot find a territory of their own to breed in go on living with their parents; their parents' territories provide enough food to support them all, and those parents are still able to raise new offspring successfully. That arrangement improves the life chances of all parties. Parent–offspring groups are stronger than pairs without offspring; as a result, even if an area becomes vacant, new settlers in it have to form groups if they are to protect themselves against the families around them. As a rule, new groups that form include at least three adult babblers.

RANK, AVOIDANCE OF INCEST, AND THE LIFE STRATEGY OF MALES AND OF FEMALES

Within each group there is a clear order of rank. Older males outrank younger males, and older females outrank younger females. All adult males outrank all females, especially females that join the group in order to breed. When fledglings come out of the nest together they fight among themselves and establish a rank order without regard to gender. As a rule this order is set by the end of their first week outside the nest, though aggressive behavior among them may still erupt during the first few months and may in rare cases lead to a switch in rank. Beyond these first few months, rank within a brood doesn't change for birds of either gender.

Babblers do not breed with their parents or with any other babbler who was a member of their group when they hatched. This rule is all but absolute: in more than twenty-five years of study, we have seen only four exceptions to it. Some babblers may live in their parents' group as long as six, or in one case eight, years without trying to breed, although they are physically able to reproduce when they are two years old. Males normally stay on in their parents' territory and take their chances there, often for life; there are territories that have seen five generations of the same male line since we began observing them and that are still going strong. If a particular group breeds enough adult males, and if the males in a neighboring group become weak, the bigger group may conquer the territory of their weaker neighbors and then divide into two groups. Of the males who eventually breed, some 50 percent do so in their birth territory; some 30 percent breed in a territory that borders it—usually after taking it over—and about 20 percent breed one territory farther out.

Females cannot mate with their fathers and brothers and so have to leave their parents' group in order to breed. Some are accepted by a neighboring group even before they are capable of reproducing, when they are as young as one year old, while others live with their parents till they are four and in rare cases even five or six. Of the females who eventually breed, a few manage to do so in their parents' territory, but only when that territory is taken over by males from elsewhere. About half the females who breed do so across the border from their parental territory, and the rest no more than three territories away.

Young males sometimes overtake their sisters in rank when they reach one or two years of age. It is hard to tell whether the males' dominance—they may now hit or shove their sisters—is a result of their own increased power or because the females choose not to assert themselves against them. After all, females have no future in their fathers' groups; they have to join other groups and breed there. Thus there is no great conflict of interest between the daughters of the group and their male siblings, and no real reason to compete with them past childhood, when they compete to receive food.

THE COMPOSITION OF GROUPS; COALITIONS OF MALES AND OF FEMALES

The simplest babbler group consists of a pair of parents and their offspring. In such a group only two individuals breed. Other groups are more complex. If a mother dies, the father, who cannot mate with his daughters, is joined by females from elsewhere. When a female joins a group, all adult males in the group are able to copulate with her; the father now competes with his sons, and with any other adult males in the group, for a chance to reproduce. The dominant male copulates most; his chances of copulating with a female when she is ovulating, and thus of becoming a father, are therefore better than those of other males. But the others can try too and sometimes succeed.

All the females in a group who lay eggs do so in a single, communal nest. Females who join groups together—two, three, or even four of them at a time—compete with one another both to lay eggs and to lay the first eggs in their group's nest; the first nestlings to hatch will have a head start on their nestmates, which means a better chance of survival and of achieving high rank within the brood.[3] Such contests are complex and fascinating. Each is a unique story, reminiscent of historical epics, Shakespearean dramas, and biblical tableaux. In one case three males were copulating with four females, all of whom laid eggs. It was impossible for anyone—human or babbler—to tell which of the fledglings in the common nest were whose offspring; it would have taken DNA analysis for us— the babblers themselves had no idea.

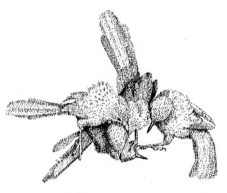

As a rule, such a partnership between females who join a group together lasts only until they bring forth one successful brood. Afterwards the dominant female becomes the only breeding female. She either kills her rivals or expels them, or they submit and become nonbreeding helpers. Sometimes one of the latter eventually casts out the dominant female—or kills her— and takes her place. Males may continue copulating with the same female over several breeding cycles and even years, but the competition for breeding privileges continues, and the losers eventually leave, are expelled, or are killed by their peers, until only one is left. The one remaining male then fathers the brood of the female who survived the struggle with her female peers; the breeding couple is now supported by a growing number of adult offspring who cannot mate with their parents or among themselves.

STRUGGLES WITHOUT AGGRESSION (WELL, ALMOST)

The most fascinating aspect of these intense internal struggles is the almost complete lack of aggression for most of their duration. Actual fighting is extremely rare, and even ritualized aggression is very unusual between two adult males or two adult females of the same group. We have seen only a few score full-fledged fights in over twenty-five years of study, in contrast to the daily border clashes with other groups, the violent chasing of nonterritorials, the fairly common aggression of males toward females, and the constant squabbles among youngsters in the first few months of their lives.

An adult fights seriously with another adult of its group only once, if ever. That one fight is all-or-nothing: the loser either is killed or flees and becomes a nonterritorial. Not surprisingly, such a fight almost always occurs between the most dominant male or female and the one next in rank: there is no point in risking everything for anything less than the top position. Such a fight is savage. It starts all of a sudden, and most observers fail to see any sign that it is imminent. It is as though all grievances and suppressed antagonisms between two brothers or between a father and son who have lived together peacefully for years burst out in this one life-and-death fight. The struggle is strictly between the two: the other members of the group observe it closely, but we have never seen any of them intervene. This is in stark contrast to battles between groups, in which all members of the group may join in an attack against a stranger or come to the aid of a group mate.

THEORIES THAT EXPLAIN ALTRUISM: GROUP-SELECTION THEORY AND ITS FAILINGS

Members of a given group help to defend their territory and to take care of the young. It might at first seem logical to assume that babblers help defend the common territory because it is this territory that ensures their survival. But each babbler could try to let others do the fighting. And why do these birds invest effort in raising offspring that are not theirs? This question, and similar issues among other species, have led in the last twenty years to a wave of studies of group-breeding birds and animals.

There are group-breeding organisms in almost all orders of animals. The list includes not only group-living birds like the babblers,[4] but also animals ranging from such mammals as lions, chimpanzees,[5] spotted hyenas,[6] African hunting dogs,[7] dwarf mongooses,[8] and naked mole-rats[9] all the way to slime-mold amebas[10] and communal bacteria.[11] Natural selection means the spread of traits that help individuals to reproduce more than others. In this contest, the only measure of an individual's success is successful offspring. It would seem that investing in another's offspring undermines one's own reproductive success and contradicts natural selection. How can one explain this contradiction?

The evolutionary model known as *group selection* once seemed to offer an explanation for the behavior of social animals like the babblers. Those who believed in group selection began with a fact that we accept: large groups can protect their own territory better than small ones, and they can take over the territories of smaller groups. But group selectionists go on to say that it is therefore in the interest of each group member to help its parents have more offspring. By helping its group grow larger, they say, the individual increases the odds that it will itself survive and reproduce.

Indeed the odds are better that members of larger groups will survive and reproduce than that members of smaller groups will. But this explanation, like all explanations based on group selection, is not stable. The flaw is inherent in the logic of group selection itself: What prevents some members of the group from becoming social parasites? Group selection, in other words, would reward individuals who exploit the willingness of others to strengthen and enlarge the group, and who save their own strength for the struggle to reproduce; thus, precisely those who work hardest for the good of the group would be penalized. And the parasites would reap another benefit: they would spare themselves the effort of caring for the offspring of others.[12]

Once it was recognized that group selection does not provide an answer to the riddle of altruism in group-living animals, two other theories were brought forward: the theory of *kin selection,*[13] which we will discuss in the next chapter, and the theory of *reciprocal altruism.*[14] The theory of reciprocal altruism is the more general one, since it does not assume the need for genetic closeness between the helper and the one being helped: one often encounters groups of unrelated individuals both among babblers and among other group-living birds and mammals.

THE THEORY OF RECIPROCAL ALTRUISM AND THE PROBLEM OF ENFORCEMENT

With his theory of reciprocal altruism, Trivers suggested that if one individual's investment in another is eventually reciprocated by an investment of the second individual in the first, or in the offspring of the first, then it makes sense for the two to invest in each other. Trivers knew that reciprocal altruism can only be maintained if there is a mechanism that ensures reciprocity. Without such an enforcement mechanism, reciprocal altruism is no different from group selection—that is, the theory would demonstrate nothing more than that a group whose members help each other is more successful than one whose members do not. That reciprocity must be guaranteed is a crucial element of Trivers's idea, but he does not specify how it can be guaranteed in the animal world.

Let's take a very simple example of the problem of enforcement, one that most of us have encountered in everyday life: littering. Those who carry a candy wrapper or an empty soda can to a trash receptacle are altruists: they take on an extra burden so as to preserve the quality of the environment for all. Those who fling away a soda can the minute it is empty are selfish individuals who save themselves some trouble at the expense of all.

It follows, then, that one who reprimands a litterer is rendering a service to the public—but may pay a price for doing so: he or she may be verbally abused or even hit by the litterer. Indeed, most people avoid reprimanding others, because it is "awkward," or because they fear being abused or attacked. Thus, those who volunteer to demand altruistic behavior are themselves altruists; it takes altruism to enforce altruistic behavior in others.

Trivers himself thought that among humans, and in apes and other animals with well-developed brains, individuals who do not reciprocate are discriminated against by other group members, and that this discrimination ensures reciprocity. The problem is that the very action group members take against those who do not reciprocate altruism is itself an effort for the benefit of the group—that is, an altruistic action. Trivers's solution to the question of altruism, then, begs the question, since his enforcement mechanism itself demands altruism. Axelrod[15] ran a computer simulation that found that punishing individuals who do not reciprocate will work, but only if any individuals who do not punish are themselves punished as well. The problem is that punishing demands effort in its own right. Axelrod concluded that the most efficient way to preserve social norms is to have professional law enforcers such as police, inspectors, and judges—who are paid to do the punishing. Of course, such professional enforcers do not exist in nature.

Trivers's theory of reciprocal altruism was accepted by the scientific community as an important explanation of altruism between nonrelated individuals. There were two reasons for this: first, there was no other theory that explained the many cases in which group members clearly invest in the good of the group, even though they have no genetic relationship to one another; and second, Trivers's theory seemed to be supported by the many instances in nature where altruism was indeed reciprocated.

The theory of reciprocal altruism led to the development of complex mathematical models, like the "tit-for-tat" model.[16] These models investigated through computer simulations under what conditions of reciprocity altruism would be desirable. With all their sophistication, none of these models managed to resolve the greatest problem posed by the theory of reciprocal altruism: How can one who does a favor for another ensure that the favor will be returned?[17]

COMPETITION OVER ALTRUISTIC ACTS IN BABBLERS

Other difficulties with the theory of reciprocal altruism emerge when we observe babblers' behavior. Babblers engage in many altruistic activities. When they eat, one of the group acts as sentry. The sentry is clearly hungry—when offered food by human observers, it often eats eagerly—but still, it stands guard when it could be feeding. Babblers give food they find to other adult members of their group. Again, they clearly feed their comrades before they are full themselves: when offered a crumb of bread right after they have fed their fellows with a similar crumb, they eat with relish—something that satiated babblers do not do. They endanger themselves by mobbing raptors and snakes. They imperil themselves by coming to the rescue of group members who get caught in a net or by a predator, or by enemy babblers during a fight. They feed the offspring of other group members and take care of them.

Not only are babblers, by all accounts, at least as altruistic as other group-breeding birds;[18] close, detailed observation shows that babblers actually *compete with one another* for the "right" to *be* altruistic. Instead of expecting their partners to return tit for tat, they attempt to *prevent* them from doing their share. The theory of reciprocal altruism cannot explain why individuals *compete* for the chance to help other members of the group, let alone why they prevent others from helping in return.

If we assume that the function of altruism is to benefit others, then it makes no difference who the altruist is, and there is no sense in *competing* for the "right" to help. On the contrary, the helper should prefer to have the benefit come from someone other than itself; the benefit to the ones being helped would be the same, and the cost to itself would be less. But if the helper benefits directly from the *act of helping*, and the benefits to others are incidental—a side effect—then it makes sense to compete for the opportunity to help. As we shall show, the helper—the "altruist"—does reap a direct gain—in prestige.

Sentinel Activities

Let's take the babblers' sentinel activities as an example. When babblers feed in the morning, one of the group often perches on a tall branch or a treetop. When this sentinel sees danger, it issues "alarm calls"[19]—which, as we have said, are directed at the approaching predator, but which also alert the group. Standing guard means exposing oneself to danger and forgoing food until the watch is over. Individual babblers may spend up to two or three hours a day standing guard. All group members take part, a fact which at first seems to support the theory of reciprocal altruism. But more detailed observation shows that the babblers are not at all interested in having others do a fair share—that is, in reciprocity.

Tirza Zahavi[20] found that higher-ranking babblers stand guard more than lower-ranking ones do, and that they interfere in the guarding done by their subordinates; they often expel their subordinates from the guard post and take their place. When a lower-ranking guard refuses to leave its post, a higher-ranking babbler who comes to replace it may shove it and even hit it. On the other hand, a babbler trying to replace a guard higher in rank than itself does so by sitting on a lower branch—sometimes on a lower branch in another tree—and waiting for the guard to leave; only then does the lower-ranked replacement go up to the dominant babbler's post (see Graph 12-1).

As a rule, the guard clearly sees the one attempting to replace it: one can observe it preening its feathers, shifting around, at times climbing to an even higher branch. The sentinel is certainly in no hurry to take advantage of the other's offer and leave

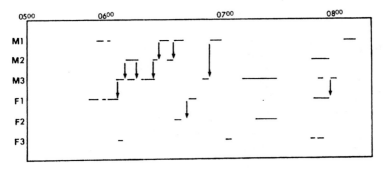

Graph 12-1. A typical sequence of sentinel activity in the morning in one group; a downward arrow shows replacement. The dominant male (M1) and the dominant female (F1) both stood guard first thing in the morning. The third male (M3) came up to replace the dominant female only after M1 went down; he was several times replaced by the beta male, M2, and once by the alpha male, M1. M2 in his turn was repeatedly replaced by M1. When the second female (F2) tried standing as a sentinel on another tree, she was immediately replaced by the top female (F1). Throughout this time, a young female (F3) stood guard intermittently; none of the others bothered to replace her.

its post. A romantic view would suggest that dominant babblers are volunteering to serve in the place of tired comrades who are younger and less experienced, but detailed observation shows that this is not the case at all. Each makes the most effort to replace the one just below it in rank, who is likely to be closest to it in age; individuals are considerably less likely to replace much younger group members.

If guarding were based on reciprocity, there would be no point in striving to do *more* guard duty than others. Even if one asserts that such competition is necessary to ensure that the group is never without a sentinel, one would still have to explain why each bird interrupts the watch of the one nearest to it in rank, rather than attempting to replace younger, more inexperienced babblers. The dominant male's interference with the guarding done by the one next to him in rank can become so intense that that second-ranked bird, the beta male, ends up guarding much less than the third in rank, with whom the top-ranking bird interferes less (see Graph 12-2).

This does not happen because of any lack of desire to guard on the part of the beta male: in fact, the beta male goes up to guard more frequently than the third in rank; the reason he ends up guarding less is that the alpha male keeps interrupting his guard duty. It so happened that in two of the groups in which detailed studies of guarding were done, the dominant males disappeared during the study. In both cases, once the dominant disappeared, the one next to him in

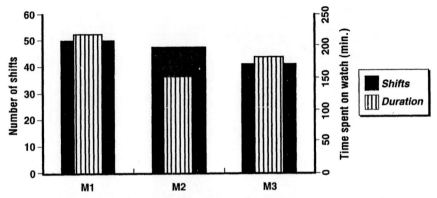

Graph 12-2. *Number of shifts and duration of watch of males M1, M2, and M3 in one group. M2 came up to stand guard more often than M3 (M2 took more shifts), but his total sentinel activity was less, because the alpha male replaced him frequently.*

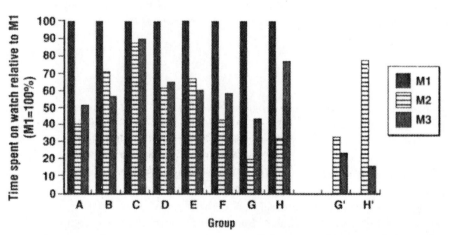

Graph 12-3. *Duration of watch for males M2 and M3 in several groups, relative to the duration of watch for the alpha male (M1); note the reversals in groups G and H after their alpha males disappeared (G' and H'). M1 = 100%. Maximum duration: 2.5 hours/day in group H. Minimum duration: 0.5 hour/day in group C.*

rank—now the top male—immediately started guarding much more than before the change (see Graph 12-3, groups G' and H'). Before the dominant male disappeared, the beta males in these groups had been guarding less than the third-ranked males; after the change, when the dominant who repressed the beta was gone, the same birds were guarding far more than the formerly third-ranked individuals (see Graph 12-3).

Feeding of Nestlings

Babblers also compete over the feeding of nestlings. Thamsi Carlisle[21] found that higher-ranking one-year-old babblers actually interfere with other babblers of the same age when those birds try to take care of the group's young. This can reach surprising extremes: Carlisle observed cases in which a group of a dozen babblers had only one surviving nestling, and four or five babblers would come to the nest with food in their beaks at the same time. Each had to  wait—with food in its beak—until all those that outranked it had finished feeding the sole nestling. Sometimes the lower-ranked birds would not even dare approach the nest (see Graph 12-4). If the only function of feeding nestlings is to nurture the nestlings, why on earth do babblers prevent other babblers from doing just that?

Feeding of Other Adults (Allofeeding)

Adult babblers, too, feed each other. Usually, within each gender such feedings go only one way: a dominant feeds its subordinate. Exceptions to this order of rank and gender are rare, and even rarer among males. The dominant female may feed the dominant male without challenging him, and other females may feed other males without dire consequences. But we have seen cases in which lower-ranked individuals who attempted to feed higher-ranking ones of the same gender were beaten up for their trouble. In at least one case, such reversed feeding was a declaration of open, aggressive revolt—an extremely rare event, as we have seen— which ended with the lower-ranked rebel being thrown out of the group.

Babblers feeding others draw attention to themselves: they emit a special trill and lift their beak above the beak of the one they are feeding. When there is a large difference in age between the feeder and the one being fed, the latter is sometimes eager to accept the food, and may even approach the feeder, making begging sounds. But in some 15 percent of cases, babblers try to avoid being fed by another bird.[22] When a guard sees a higher-ranking comrade approaching it with food, to feed it and replace it on sentinel duty, it may sacrifice its guard post to avoid being fed. In other cases, the sentinel may close its beak tightly and refuse to accept the food being offered, even though it is hungry; it eagerly accepts a dry crumb of bread from us immediately after refusing a juicy insect from another babbler. If the aim of allofeeding is to support weaker group members, why do babblers try to feed individuals who do not want to be fed, and why do the latter resist?

In their first months of life, fledgling babblers are still aggressive and hit each

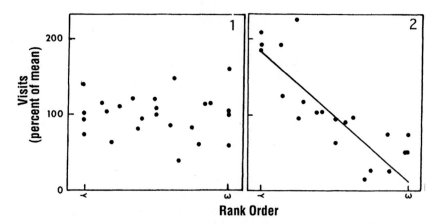

Graph 12-4. Visits to nests by yearling helpers, as percentage of mean visits. Data and graphs by Thamsi Carlisle (Carlisle and Zahavi, 1986).

Left: Visits to unattended nests. Summary of 598 visits to 4 nests by 3–9 yearling helpers. When there was no other bird at the nest, there was no apparent relationship between the rate of visits to the nest and the helper's rank.

Right: Visits to nests where the incoming helper had to replace another at the nest. Summary of 187 visits to 3 nests by 3–9 yearling helpers. The solid line indicates that the lower the rank of the incoming bird, the less frequently it came to the nest when another helper was there.

other frequently. They may even switch ranks during their first year of life. Thamsi Carlisle[23] recorded 94 cases of fledglings and yearlings feeding each other. In 86 cases a yearling offered food to its subordinate. In 25 of those—well over a quarter—the subordinate refused the offered food or only tasted it. In 7 of these 25 cases of refusal, the would-be feeder chased the subordinate and sometimes pecked at it. Only in 8 cases did a subordinate yearling offer food to one ranking above it. In fully 6 out of these 8 instances, the dominant yearling refused to eat and attacked the would-be feeder. In Carlisle's research, most feeding was done with breadcrumbs given to the birds by the observer. More recently, Amir Kalishov studied allofeeding among adult babblers without offering them any food; his findings were similar to Carlisle's.[24]

In one case in Carlisle's study, an eight-month-old female tried to feed her ten-month-old sister; the latter stood up, snatched the food out of her younger sister's beak, forced her to crouch like one begging for food, and then stuffed the food down her throat. Once the younger bird had swallowed the food, the dominant sister pecked her until she fled. The frustrated younger sister then took another crumb and went to feed her subordinate younger brother, who was hunting quietly for food some thirty feet away.

In another case—one of the two instances in Carlisle's study in which the gift of food was accepted by one higher in rank than the feeder—the dominant fledgling accepted the food but then attacked the feeder and kicked it in the face, as if to say "I'll take this food, but I still outrank you." Far from rewarding altruistic behavior and returning tit for tat, babblers who are offered food by their subordinates actually punish them for making the offer. If the function of feeding others were to benefit the ones being fed, why would the recipients of the benefit punish their benefactors? Babblers behave as though it is the act of giving, rather than the benefit given, that matters.

Mobbing

Babblers' mobbing is generally similar to that of other vertebrates. A babbler who notices a snake or a raptor will sound a tzwick and then make barking and trilling sounds. This alerts the other members of the group, who arrive and approach the raptor. When a babbler comes within four to six feet of the raptor, it spreads its tail and lifts its wings, fanning them out. It holds that stance for several seconds; it may turn sideways and circle the raptor, tail and wings still spread. These displays last a minute or two, and the babblers' movements change according to the degree of danger. Zahava Carmeli, who stud-ied babblers' mobbing of snakes, found that the babblers were well aware of the risk: she determined that on average mobbing babblers kept a distance of 10 to 15 inches from a snake's tail, but they kept some 20 inches from its head, which would indicate that the birds knew full well which end of the snake posed a danger.[25] If the babblers attacked the snake at all, they aimed at its head. Mobbers' behavior varies a great deal: some dive headlong into danger, others hold back. Different babblers stop at different distances from the predator.

The tendency of a babbler to mob correlates to the social makeup of its group. Avner Anava studied babblers' mobbing of raptors, often using stuffed models. Two of the groups he studied comprised one adult male living with several females and year-old fledglings; in these groups, the dominant male did not participate much in mobbing (see Graph 12-5, group HAK). In groups with several adult male members, all those adult males participated much more. In these groups, the dominant male always mobbed longer than the other males, came closer to the raptor, stood guard during mobbing more than others, and actively disrupted the mobbing activities of the other adult males (see Graphs 12-5 and 12-6, groups ZEH and MZR).

The activities of the second- and third-ranking males depended mostly on the relationships between them and the first male. When such males accepted unquestioningly the position of the dominant male—for example, if they hardly made any

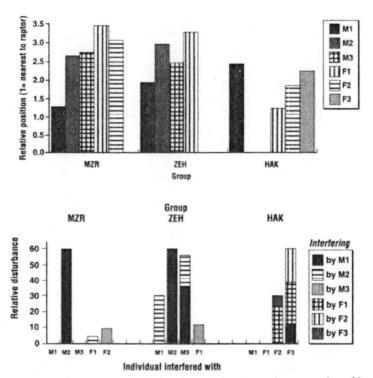

Graphs 12-5 and 12-6. Competition over mobbing and interference with mobbing. Data from Anava, 1992. At the time, group MZR comprised a father (M1), a mother (F1), an adult son of the father but not of the mother (M2), and a son and a daughter of both father and mother (M3 and F2). Group ZEH comprised three sibling males and a female who was not related to any of them. M1 and M2 were from the same brood, and during the course of the study there was intense conflict between them; eventually, M1 disappeared—probably ousted by M2, who then took his place. Group HAK was composed at the time of a father (M1), a mother (F1), their daughter (F2), and a stranger female who had recently joined the group (F3).

Graph 12-5 (top): Position relative to the raptor: nearest = 1, farthest away = 4. In the MZR and ZEH groups, the dominant males were almost always the ones nearest the raptor, and the dominant females, who were the only adult females in each group, stayed farthest from the raptor. In group HAK, which comprised one adult male and three adult females, the dominant male stayed farthest from the raptor, and the dominant female came nearest.

Graph 12-6 (bottom): Interference during mobbing—who interferes with whom. In the MZR group, no one interfered with the dominant male (M1). The dominant male himself interfered a great deal in the mobbing activity of M2. No one interfered with M3, but M2 interfered with the dominant female (F1), and M3 interfered with his sister, F2. In the ZEH group, M1 interfered a great deal with M2, but M2 also interfered with M1. Both M1 and M2 interfered with M3—M1 more than M2. M3 interfered with the female only. In the HAK group, M1 did not interfere with F1 but did interfere somewhat with the stranger female, F3. None of the other females interfered with F1. Both F1 and F3 interfered with F2, and both F1 and F2 interfered a great deal with mobbing by F3.

attempt to copulate with the breeding female at laying time—they mobbed less than he did: they stayed farther away from the predator, mobbed for shorter periods, stood guard less during mobbing, and sounded fewer and less orderly trills (see Graph 12-5, group MZR).

At the other extreme were two groups in which there was a more intense conflict between the dominant and the beta males over copulation. In both cases, the first and second male participated almost equally in mobbing, and the first interfered a great deal with the mobbing of the second. The interference was such that in one group the beta male ended up mobbing even less than the third in rank. One of these groups was the only one in which the beta male dared interfere with the mobbing of the first (see Graph 12-6, group ZEH). In both groups the first male disappeared eventually, probably kicked out by the second male who then took his place: in one case this happened during the study, in the other, right after it ended. The differences in mobbing between females also correlated with competition among them.

A dominant babbler's interference in the mobbing activities of its subordinate was usually subtle; the dominant might come and sit right next to the lower-ranked mobber, preen the subordinate's feathers, or push it lightly. In such cases, the subordinate would cease for a time to utter the trilling sounds babblers make during mobbing and would crouch in place or back off. Sometimes the dominant was rougher: it might flank its rival abruptly, cross the rival's path, or land near or in front of it. The other would react with a frightened jump accompanied by a tzwick and stop mobbing temporarily, or even stop altogether. Again, the dominant male intervened in mobbing by the second-ranked male, who is usually an experienced adult, far more than in the mobbing of youngsters. In other words, it is not to protect inexperienced group members that the dominant bird interferes.

Incidentally, mobbing crows have also been found to compete for prestige. Slagsvold[26] showed that higher-ranking individuals mobbed more than others and even attacked and chased away lower-ranking individuals during mobbing. He suggested, as we have, that the higher-ranking crows advertise in this way their control over the flock.

ALTRUISM AS A SUBSTITUTE FOR THREATS

In short, babblers compete to perform altruistic acts, and often they even attempt to prevent their subordinate comrades from acting for the good of the group.

Evidently, each babbler cares mainly about *providing* the benefit, rather than about the actual benefits the group receives. If we want to understand the babblers' altruistic behavior, then we have to determine what the giver—the "altruist"—gains from the act of giving, rather than what the group as a whole gains.

It is important to remember that group members of the same gender compete with each other over the utmost biological need—the chance to reproduce. In spite of this rivalry, they hardly ever fight one another. In the vast majority of cases, conflicts within the group are settled by means of the competition over altruism.

How can altruism replace physical aggression? This brings us back to the substitution of threats for actual fighting (see chapter 3, on rivals). Because of the danger and cost inherent in fighting, to both loser and winner, other ways of resolving conflicts evolved. Actual, all-out physical struggle was replaced by threats that showed with a reasonable degree of reliability how likely the threatener was to win an actual fight.

But threats have their own price. When threats are not effective, the threatener does have to fight—or forfeit. The stakes are highest when an animal issues a threat in the presence of witnesses; when an individual's threats go unheeded, the individual may fail to deter not only the rival but the witnesses as well. Thus, the presence of witnesses tends to make threateners less likely to compromise. Moreover, witnesses who see a contestant injured or spot some weakness in one or the other may seize the opportunity and take over. This is especially true among highly social animals like babblers, who are always together. Any threat not followed up by action, any unsettled conflict, any failure of a dominant member is noticed by the entire group—comrades who are also potential rivals.

Yet conflicts do occur and demand resolution—so a substitute for threatening—which itself evolved as a substitute for fighting—is highly desirable. Actions that are not direct threats but are closely related to an individual's ability to win a struggle can take the place of threats. By showing off how much they can invest in standing guard, in feeding their comrades, in taking risks, and in other altruistic acts, babblers show off their ability to win in a fight and their desirability as group-

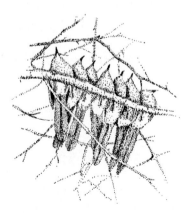

mates. The altruist's investment in the altruistic act offers a reliable, concrete index of that individual's ability.

Altruistic acts have another benefit: they prove the giver's interest in the receiver, even while they proclaim the giver's dominance. By investing in the good of the group, the dominant male shows off both his superiority and his willingness to give to his subordinates. This makes these subordinates less likely to leave and thus helps the dominant bird remain at the head of a large, strong group.

Other group members have good reason to pay attention when an individual acts for the group's benefit: what they learn about that individual helps them avoid a high-risk fight against a rival stronger than they are. All members of the group must constantly be assessing their prospects. After all, each bird has to decide whether to stay on in the group as a subordinate, leave and try its luck elsewhere, or risk everything in a fight for the top position. Altruistic acts that benefit the group help each babbler answer these very questions.

RANK AND PRESTIGE

Rank within the group is very important in the life of babblers. Every social activity, from fighting other groups to dancing and mobbing, is affected by and reflects social rank. Babblers sleep clumped together along a tree branch: the dominant male always sleeps at one end, and the second-ranked male usually at the other end. More dominant birds usually stand guard more than others. They often take over from their subordinates, while their subordinates avoid trying to take over from them. Each babbler feeds its subordinates, while the latter very rarely feed their superiors. Subordinates sometimes make a special sound when a more dominant bird passes by them. Dominants can preen their subordinates' backs and tails as well as preening them elsewhere; subordinates preen babblers that outrank them almost exclusively in areas other than the back or tail. Why is rank expressed so vigorously when for all practical purposes it does not change as long as a babbler stays in its original group?

In time we discovered that the relationship between the first and second males is not the same in all groups. It depends a great deal on the particular group's makeup. Even when the two are brothers, as is often the case, this relationship varies a great deal from group to group. At one end of the spectrum, the dominant male is an old, experienced babbler and the second is five or six years younger. In such a case, the second never dares feed the first during the dominant's guard duty, always makes the submissive sound when the dominant approaches, and opens his beak to ask the superior for food. At the other extreme, the second-ranked bird is of the same brood as the first and became his subordinate as a consequence of childhood squabbles. In that case, the second stands guard a great deal, the first feeds him only infrequently and is often refused; on rare occasions the second will even feed the first. The social rank is the same in both cases—one is clearly the first and the other clearly the second in rank—but the difference in the degree of control is very significant.

To reflect these differences in the quality of relationships within groups, we use the term *prestige,* which for us is different from and complementary to rank. (In previous writings we used *social status.*[27]) Prestige is the respect accorded an individual by others. In social systems where the strong have little interest in the well-being of others, prestige can be won by unprovoked aggression, whether physical attack or threat. Such aggression is so common that social ranking is also called "pecking order."[28] Prestige can also be acquired by "wasteful" showing off; the peacock's tail, the deer's antlers, and the bowers built by bowerbirds are examples of showing off that is neither harmful nor useful to others. Babblers and other altruistic animals gain prestige by investing in their fellow group members' well-being. Their partners benefit, but the reason for their altruism lies elsewhere—in the real testimony to their ability that it provides.

Social rank is easy to discern; prestige, on the other hand, is complex and harder to measure precisely. Prestige reflects the *degree* of a superior individual's dominance, as recognized by subordinate members of the group. In other words, prestige is gauged by others. The dominant may claim prestige, but for the prestige to be real it has to be accepted by subordinates, and it is this acceptance that actually determines an individual's prestige.

Prestige has real value: a dominant male with higher prestige can get with ease what a dominant male with less prestige gets only with great effort, or not at all. Babblers make constant, unceasing efforts to stress their ability, not to rise in rank—an adult babbler can only "promote" itself by killing its superior or chasing it from the group, or by leaving the group itself—but to persuade their comrades to recognize that ability and grant them prestige. A babbler who can stand guard longer than its comrades, give them part of its food, approach a raptor, take the risk of sleeping at the exposed end of the row—and can also prevent others from doing such deeds—proves daily to its comrades its superiority over them. By doing so, that individual increases its prestige and has an easier time exerting control.

Greater prestige means that subordinates will avoid getting into a fight for the higher position. Greater prestige convinces the female to stay faithful to the dominant male and not to mate with his subordinates. In short, a babbler's altruistic acts are an investment in its prestige, and that prestige has real, concrete value. The difficulty—the cost, or handicap—of this altruistic advertising, whether food sacrificed or danger incurred, is what makes it a reliable indicator of the ability of the advertiser.

Subordinates have their own prestige, which is shown by their ability to resist their dominants and by their superiority over those below them. Babblers replaced during guard duty or fed by one more dominant than they often try to regain  prestige as best they can by replacing a lower-ranking guard, or by attacking a bird of a different species, like a bulbul or a blackstart. The frequency of these behaviors shows clearly both that individual babblers are very aware of their own prestige and that many of their actions, especially altruistic actions, are done to gain prestige.

We also find abundant evidence that other group members are well aware of their comrades' claims to prestige. Subordinate males often approach a female or a lower-ranked male who is standing guard, feed it, and take its place as a sentinel; in some groups, this almost always provokes a reaction from the dominant male, who hurries to feed the subordinate guard and replace him. Sometimes the dominant male stands guard only for a moment before leaving the guardpost—as if to say that his only intention was to make his rival stand down and assert his superiority, which may have suffered slightly from the subordinate's display of prestige.

"SHYNESS" OVER COPULATION AS A TEST OF MALE PRESTIGE

Babblers' copulation rituals are tests and demonstrations of the dominant male's prestige. Most animals copulate freely in the presence of other members of their species. Not so babblers. Even though babblers spend most of their lives within sight of their fellow group members, they do not copulate in their presence: they hide under or behind bushes and copulate in privacy.

The other group members know what the pair are doing in their hideout, since the invitation to the act is often made in public: a male inviting copulation approaches the female with a bit of food or a twig in his beak, and instead of feeding her, lifts up his gift and moves away from the group. If the female wishes to copulate, she follows. If she does not follow, the male may beat her and chase her, but this is usually ritualized, show-off aggression only. We have never yet seen a case of a female being thrown out of a group because she refused to copulate with the dominant male, though we often see females refusing and being harassed.

Sometimes one of the other males follows the pair. In such cases the dominant male stops, turns and approaches the subordinate, "reprimanding" him vocally and sometimes circling him. In most cases the subordinate male halts, acknowledges his inferior status, and returns to the group. But in a significant minority of cases he is not deterred by the dominant male's warnings. In some of the groups we have observed, the second male continued following the pair, preventing the dominant male from copulating for a number of days. Such a conflict sometimes ends in a real fight in which the defeated party is killed or thrown out of the group; when it does not, it proves to the female that this particular dominant male does not wish to or does not dare fight his second in rank, presumably because he cannot afford to lose him as a group member. The ability of the second male to interfere is a clear measure of his prestige. Indeed, in such cases we often find that the female copulates later in her fertile period with the second male as well as with the first.[29]

Each male's reproductive chances depend on cooperation from the female. The female needs to gauge as precisely as she can the current prestige of her suitors—no simple feat, since the competition over breeding with her is likely to be most intense within days of her arrival in the group. At that point, the new female does not yet know the group well; moreover, the senior subordinate males in the group are not her offspring, and thus all of them are eligible to copulate with her. The new female can learn the males' ranks quickly and easily, but rank alone does not tell her the degree of control the dominant male has over his subordinates. How much does he need them? How much do they recognize his dominance?

The new female has to split her male suitors' chance to breed according to the balance of power among them, but how can she find out what that balance of power is if they live together in peace, with no overt aggression whatsoever? If the female makes an error in granting or withholding her consent, she can end up with a powerful, dissatisfied male groupmate who will break her eggs. We have observed several cases of males breaking eggs in the nest under such circumstances. The main victim is the female, who invests more in producing eggs than do her consorts.

The "shyness" that prevents the female from copulating in the presence of other babblers helps her determine the prestige of each male—or, to be more precise, it allows her to see how much prestige each male is accorded by the others. This enables her to share her favors according to the relative importance of her partners. The dominant male's prestige is demonstrated when other males refrain from following him and the female even at a distance. Babblers who let the dominant male retire with the female without even trying to observe them from a distance show their acceptance of the dominant male's superior prestige. By contrast, a male who follows with impunity displays to the female his own high prestige as a subordinate.

A few days before egg-laying, the dominant male starts "guarding" the female.[30] He follows the breeding female closely and keeps her in sight; at whatever time of day, he is nearer to her than is any other group member. When another male approaches her, the dominant male does so too. In most cases his approach is enough to cause the subordinate to make the submissive sound or show his discomfort by fluffing his feathers and preening himself.

The dominant male always copulates with the female more than any of the other males do. But other males, too, may copulate with the female if they are not her offspring and if she cooperates. If the female wishes to copulate with another male, she can always find the one or two minutes needed when the dominant male is absorbed in something else, such as a border conflict or nest-building. In some groups, subordinate males manage to copulate with the female even on her fertile days and even in the mornings, which seem to be the best time for fertilization. Yoel Perl found that in some cases the dominant stops guarding the female even while she is still fertile, before the ovulation of the fourth egg; the second in rank then copulates with her and presumably fertilizes that egg.[31] In other groups, males of similar rank do not copulate at all. An individual's reproductive chances, then, cannot be predicted by rank alone.

When the dominant male allows his subordinates to mate with the female, he does so because he needs them in the group and must therefore make concessions to them. We think that the position of each group member depends both on the balance of power within the group and on the group's neighbors. Some groups have strong neighbors and need more and better fighters for protection, while others have weak neighbors. Some dominant males are strong and do not need much assistance in fighting; others depend more on their subordinates.

Another factor that can affect the balance of power within a group is that the more partners the dominant male has, the less dependent he is on any single one of them. And a dominant male with adult sons by the current dominant female— who do not mate with their mother—depends less on the assistance of other males who might compete with him over copulation. All these factors affect cooperation among males. Every group of babblers has one male who outranks all the others, but some dominant males let their comrades copulate, while others threaten them or even peck at them when they try. Clearly, rank is only part of the picture; prestige, too, is a crucial element.

REASONS FOR AND CONSEQUENCES OF LIVING IN GROUPS

The imperative to live in groups is powerful among babblers and is reinforced in various ways. Babbler groups make it difficult for solitary pairs to succeed on their own. When a pair tries to breed near a large group, their neighbors give them a hard time: we have observed several raids that caused a breeding pair to desert their nest or led to the killing of their fledglings. Even after years of drought, which make it impossible for babblers to raise young and leave some territories empty, subordinate adult babblers in neighboring territories do not usually hasten to settle on their own in the unoccupied lands. Instead, they often stay in their original territory and compete with their brothers for the chance to reproduce. Presumably, being the second male, or even the third or fourth, in a large, secure group with a good territory is better than being the dominant in a marginal territory, or trying to live without supporting subordinates. It seems that male babblers prefer to split off and try to form their own group only when they have at least one helper to accompany them, and when two or more females from other groups are available to join them.

It may well be that, over time, some organisms develop characteristics that help them live efficiently in groups and lose others that might better enable them to live singly or in pairs but are detrimental in a group setting. For example, in an unfamiliar situation, a group member can afford to refrain from action and observe its experienced comrades, rather than experimenting and possibly failing. Babblers indeed benefit from an extended period of learning, during which they acquire both group-living skills and abilities that enable them to find food, avoid predators, and deal with strange babblers. Likewise, in a group setting, it pays to be inclined to compromise rather than quick to fight over one's rights; we have seen that it is exactly this temperament that characterizes babblers.

Adaptations such as these may compensate for the disadvantages of group life; this may explain why the scores of species of babblers in the Indian subcontinent all live in groups, although they live in many different habitats, which are also occupied successfully by solitary songbirds.[32] After all, some adaptations are likely to be irreversible, since they involve a complex set of interconnected changes. Reptiles and mammals who left the land and returned to the water continue breathing through their lungs, even though in their new habitat it would benefit them to breathe through gills. If babblers could all at once regain the characteristics they lost in adapting to group life, they might do better as pairs in some new environments; as it is, they do better living in groups.

Some species of birds live in groups or in pairs depending on conditions. Seychelles warblers, which usually live in groups, were transferred in one case to an island without warblers. They formed pairs, nested, and increased in number until all suitable territories were occupied, and then they went back to group living.[33] For these warblers, group living seems to be a relatively new development: unlike the babblers, they don't seem to have group-living relatives.

Group living has developed in different species for different reasons, and there are many different kinds of collaboration. Some provide an edge in the competition for resources, whether with other species or with other groups of the same species. Others afford better protection against predators or adverse environmental factors. Among some birds, like the scrub jays of Arizona, coalitions are limited to parents and adult offspring; in such species the adult offspring do not share in reproduction.[34]

Some other jays, some African starlings, and long-tail tits form coalitions to protect their territory. Each group is made up of several pairs, but each pair has its own nest. Individuals whose nest fails or who do not find a mate assist the other pairs at their nests and help protect the common territory, but they do not themselves reproduce. The groove-billed anis of Central America form new coalitions every year, with several pairs sharing a single nest.[35] Simpler coalitions may consist of several females who lay eggs in the nest of a single male, as in the case of ostriches, or of several males who help a single female raise her young, as do Galapagos hawks and the harriers of Scotland.[36]

PRESTIGE AND THE EVOLUTION OF ALTRUISM: ALTRUISM AS A HANDICAP

In all cooperative animals, whatever the primary reason for forming the group, conflicts among partners are settled to one degree or another by the competition for prestige. An individual's prestige is reflected in—indeed, defined by—the respect accorded him or her by others. This respect allows the dominant male to keep others as his partners and to control his group without punishing them. Greater prestige ensures an individual a bigger share of the partnership's "gains"— that is, a better chance to reproduce successfully. Increased prestige for one partner means a loss in prestige for another. In other words, it is a zero-sum game within the group.

Scientists increasingly recognize that many altruistic activities observed among animals cannot be explained by the theories of kin selection and of reciprocal altruism. Some have suggested recently that even though altruistic behavior is not reciprocated by others directly, it is rewarded by "indirect reciprocity": others "reciprocate" by according status to the altruists.[37] But this concept stretches the definition of reciprocity to the point of meaninglessness. That the altruist gains in status, or prestige, is precisely what we have been suggesting for years. And once one accepts that altruistic activities bring those who perform them a gain in status, then no further explanation for altruism is needed.

Both the competition over altruism and the effort by others to avoid receiving benefits handed out by altruists run counter to the theory of "reciprocal altruism." Indeed, one can even view altruistic acts as implied or surrogate threats, since the prestige the altruist gains allows him or her to achieve what other animals gain by threats. Observation suggests that altruism is a good indicator of an individual's quality, both as a partner and as a rival—and thus it can substitute for overt threats.

Needless to say, a group whose members compete for prestige by demonstrating their altruism will be better equipped to compete against other groups than a coalition whose members vie for prestige by harassing those of lower rank or engaging in wasteful display. But it is important to remember that what motivates the individual is not the benefit to the group but the enhancement of his or her own prestige; the benefit to the group is a side effect. If individuals did not gain directly in this way from their investment in the good of the group, there would have been no stable foundation for the development of altruism. Logically, we assume that if circumstances change in the future and babblers or other social animals find it more beneficial to gain prestige by wasteful display, intragroup fighting, or threats, altruism will disappear in spite of its benefits to the group.

Once one views altruistic acts as signals that show the ability and the intentions of the altruist, then altruism no longer poses any evolutionary enigma. The investment in the altruistic behavior is the factor that ensures the reliability of the signal.[38] To the altruist, the cost—the waste—of altruism is no different than the waste of, say, growing and carrying a large, heavy, decorated tail like the peacock, or building an elaborate bower like the bowerbird. The cost of these and any other signals is the handicap that ensures that they are reliable—that the signaler is indeed what he or she claims to be. Thus there is no need to look for a special evolutionary mechanism to explain the evolution of altruism. On the contrary, we believe that our explanation of altruism as a signal holds true not only for babblers and other birds, but also for mammals,[39] including humans; for social insects; and even for one-celled organisms.

CHAPTER 13

THE SOCIAL INSECTS: WHY HELP THE QUEEN?

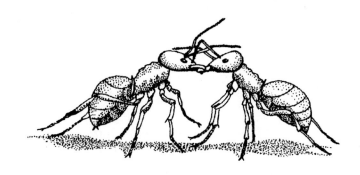

In natural selection, "success" means success in reproduction. Out of tens of thousands of seedlings on a forest floor, only a handful survive to become fruit-bearing trees; out of thousands of fish eggs, few develop into reproducing adults. Most individuals die before they ever have a chance to reproduce. This tiny survival rate does not conflict with the theory of natural selection but rather displays it in all its potency: failed seedlings, or fish that have not reached maturity, die and disappear.

Among bees, too, only one in tens of thousands in a hive may bring forth descendants. That in itself would not cause a raised eyebrow. But seedlings or fish eggs simply disappear; the female workers in a hive—that is, the eventual losers in reproductive competition—rather than focus all their efforts on trying to survive and reproduce themselves, actually expend effort to help the queen reproduce. How could such behavior evolve by natural selection?

This question has been seen in different lights at different periods. For Darwin, the main question was a technical one: How can a queen, so different from workers, pass on to her offspring the characteristics of workers? After all, both the body structure and behavior of a worker bee is very distinct from those of a queen bee, and among ants and termites, the different types of workers vary greatly not only from the queen but also from one another. Darwin did not see any difficulty in the existence of sterile individuals who assist the group: he simply assumed that a queen whose daughters were better workers had an advantage over a queen whose daughters were less well-adapted to their task.[1] But worker bees, wasps, and ants of many species are not inherently sterile, and some actually do reproduce. So researchers now pose a different question as the central issue in the evolution of the social insects: Why do the workers, who retain the ability to reproduce, work for the hive, rather than devoting their efforts to whatever chance they may have to produce their own offspring?

THE EVOLUTION OF SOCIAL STRUCTURES IN THE SOCIAL INSECTS

Conditions that Favor Collaboration: Food Storage and Helpless Offspring

First, we need to find out why the social hymenoptera—bees, wasps, and ants—band together at all. The answer may be found in their particular strategy for raising offspring. Unlike the larvae of butterflies or grasshoppers, who can move and seek food by themselves and whose success depends on their own ability to find food and avoid enemies, the larvae of social hymenoptera are fed by their mother or by other caregivers as they grow. This allows them to use food gathered over much greater distances, far beyond the reach of such larvae who get to their food by crawling—thus, more food is available to them. It also saves them the need to move: in fact, they are eyeless and legless.

An offspring's dependence on food supplied by the adult does not necessarily cause the development of a social structure, but it was a major factor in the development of social organizations in hymenoptera. Both the stationary larvae and the food gathered for them attract parasites, predators, and same-species robbers. These dangers make it desirable to build a nest for the eggs and the larvae, and to protect that nest. Partnerships can provide better protection: one partner can seek food and leave the other guarding the nest. In evolutionary history, this fact was the basis of collaborations—but it also made it possible for stronger partners to exploit weaker ones.

The larvae's dependence on food supplied by adults allows the latter to control each larva's size and destiny. The eggs from which queens and workers develop are the same. In order to develop into workers rather than queens, larvae have to be repressed by their caregivers—through undernourishment or pheromones or with physical force. Future workers get small amounts of inferior food; larvae that are designated by the adults to develop into queens receive large quantities of superior food—like the honeybee's royal jelly. The caregivers thus divide their resources unequally and deprive some of the offspring in a way that lessens the latter's chances to reproduce. Parents exploit some of their offspring by making them into helpers who will assist them or their other offspring to reproduce.

The theory of "parental manipulation," as it was called by Alexander,[2] assumes that the parent is concerned with the sum total of its reproductive success—but not necessarily with the fate of one or another offspring.[3] This explains the benefit to the parent in exploiting some of its offspring, but it does not explain the other side of the question: Why does the shortchanged offspring help its parent? Alexander assumed that the theories of kin selection and of reciprocal altruism explain adequately why the repressed workers assist their reproducing kin. We disagree.

The Haplodiploid Mechanism of Gender Determination

Another factor played a part in the evolution of insect social systems: the haplodiploid mechanism, a system that determines gender in hymenoptera (as well as in some other insect groups, but not in termites). Males come out of unfertilized eggs and are haploid—that is, they have only one set of chromosomes, inherited from their mother. Females hatch out of fertilized eggs and have a diploid (double) set of chromosomes, half from the mother and half from the father. A fertilized female—a queen—keeps the sperm she receives in a special receptacle near her oviduct and can lay either fertilized eggs that hatch females or unfertilized eggs that hatch males. Unfertilized females—workers—can only lay unfertilized eggs that hatch males.

Selection Through Queens Only, or Selection Through Workers Too?

If workers were completely sterile, then workers' traits could change only through a mutation in the queen that she bequeathed to her daughters, both workers, who would exhibit the changed trait, and new queens, who would pass the trait on to their daughters. In that case, a hive could evolve in the same way as a multicellular body, with the queen as the organ of reproduction and the workers as the other body parts. But in fact there are a great many reported cases of female workers of all orders of the social hymenoptera—bees, wasps, and ants—laying eggs. Sometimes, indeed, a large percentage of the males come out of eggs laid by workers.[4] The fact that many workers can and do lay male eggs brings forth another possibility. Is it possible that by laboring for the hive, workers are actually doing the best they can to raise their own sons?

This would open up a completely different avenue for the evolution of the hive's social structure. Let's assume that by working for the hive, a worker could increase her own chances to have successful sons. Then any mutation that improves the worker's ability to labor for the group will enable her both to produce and raise more sons and to pass the mutation on through them to the following generations of queens, workers, and drones (males).

The fact that caretakers can manipulate female larvae and raise workers or queens at their discretion on the one hand, and the workers' ability to lay male eggs on the other, made possible the creation of asymmetrical partnerships between large, fertilized females—queens—and unfertilized, mostly smaller, females—workers. A worker cannot establish her own colony, since she cannot lay fertilized eggs and bring forth daughters who would assist her in guarding her colony and taking care of her offspring. But she can stay in a colony, assist the queen to raise more workers, and try to lay eggs that will hatch males. That colony may be either her birth colony or another; among stingless bees there are even cases in which queens join colonies established by workers.

In our opinion, the chance the worker has to reproduce within the colony is the cement that permitted the creation of large, stable partnerships, encompassing thousands of individuals, in which the reproductive success of the individual workers depends on the success of the queen. The fact that the queen is able to raise daughters smaller and weaker than herself makes it possible for her to exploit them; it is the inequality between queen and workers that limits the workers' options and makes the asymmetrical partnership so stable.[5]

HOW INSECT COLONIES FORM

Colonies of hymenoptera are formed mainly in three ways. First, by coalitions of queens of the same age; these are usually sisters or are otherwise related, but sometimes they are strangers. Second, by coalitions of a mother and her daughters. Third, by the breakup of an existing colony into several daughter colonies. To illustrate the workings of such partnerships, let's examine several cases.

In many species of wasps and ants, several queens join together to form a new colony.[6] In the beginning, the partners need each other, and often more than one queen lays eggs. But once smaller worker females hatch, the top queen can afford to get rid of her partner–rivals. And indeed, at that point the founding queens usually start fighting among themselves, and the winner kills or expels her partners or enslaves them. In most cases the colony ends up with one queen only, who goes on laying eggs, while her daughters and the daughters of her former partners take care of the offspring. These workers, who are not fertilized, cannot compete with the queen effectively since they are smaller, and since she is the only one who can lay fertilized eggs to create more workers.

In the temperate regions there are many species of hymenoptera that form annual colonies in the spring, which die out with the onset of winter. This means that in the spring there are no large colonies, and a single queen can establish a new nest on her own. For example, young queens of the wasp *Vespa orientalis* and its relative *Vespula germanica*[7] hatch in the fall and live for a while in their birth colony. At that time, they do nothing but eat and rest. Later, when the colony nears the end of its life, each queen goes on a mating flight, then passes the winter in hibernation in a crack in the soil, often with siblings. In spring she wakes up and starts building a nest, laying eggs, and gathering food for her offspring.

The first offspring of such queens are females. They cannot mate, since there are no males at that time of the year; so they function as workers, helping the queen to raise additional offspring throughout the summer. Late in the summer things change. The queen starts laying unfertilized eggs—male eggs—and the workers start giving some of the queen's fertilized eggs large quantities of superior food; these eggs develop into the next year's queens. At the same time, some of the most dominant workers manage to lay eggs: each takes over several cells, which she protects, and in which she lays unfertilized (male) eggs. Thus, some of the males that hatch in the colony are offspring of the workers, not the queen. The laying workers are very aggressive toward each other and toward the queen. They eat eggs laid by their rivals and by the queen and destroy the cells of their rivals.

Why do these workers stay in the colony? For two reasons. First, a worker, being small, has no prospects on her own against the other colonies around, each with a large queen and a number of workers. The workers are small because their mother has raised them to be small, and the first workers are particularly small because their mother has raised them on her own and had a hard time bringing in enough food for her larvae. Second, a worker cannot lay female eggs to raise daughter workers who would help her protect and take care of her nest. Thus, the only way a worker that hatches in the spring can have viable offspring is to survive in an established colony until the fall—the time for raising queens and males— and then attempt to lay male eggs. Since she has to stay in a colony, she may well choose to stay in the colony she is familiar with and hope conditions will allow her to bring up successful sons when the time comes. But she does have the option of moving to another colony, and in fact, many workers do move from colony to colony,[8] presumably seeking a place where they will have a better chance to reproduce.

Discussions about colonies usually center on conflicts of interest between workers and queen, but in fact each worker has her own individual interests; a large part of the life of a colony makes no sense unless conflicts of interest among individual workers are taken into account.[9] There is an order of rank among the workers. If the top worker were to kill the queen and take her place, that worker might be able to lay her own unfertilized male eggs freely, but the other workers would be repressed by that upstart worker. And their colony would be weaker; only the fertilized queen can strengthen the colony by laying female eggs and creating more workers.

Each worker's chance to have sons, and those sons' chances, depend on the well-being of the colony, and the colony can be strengthened only by the queen. So it is better for most workers to have a queen at the head of the colony than to have another worker in that position. Each worker's chance to reproduce success- fully conflicts more with that of the other workers than with that of the queen. To get the best shot at having successful sons later, then, younger workers have to cooperate with the queen and suppress egg-laying by the highest-ranking workers. No wonder that all summer long the colony grows by raising more workers: the workers themselves prevent the raising of princesses (young queens) and males, and they suppress and kill workers who try to lay eggs.

In the fall, the case is different: the colony is nearing the end of its life. The time is ripe to raise new queens and males. The queen begins laying unfertilized male eggs as well as fertilized female eggs, and the workers start giving some of the queen's fertilized eggs the treatment that will make them into future queens. At this season a struggle often breaks out over the laying of male eggs, both be- tween workers and queen and among the workers. The winners among the workers manage to lay some of the colony's male eggs—sometimes even most of them. The workers who do not manage to lay male eggs help the queen and their egg-laying comrades raise their young.

A similar logic applies when existing colonies break up into new ones. To withstand neighboring colonies, a colony of wasps has to avoid breakups, which would spread its strength too thin. In that situation, it is in the interest of work- ers to suppress the raising of princesses and the laying of male eggs by other workers—since, when grown, these would swarm and split the colony. But once the colony is large enough to become two viable ones, a split is in the interest of both queen and workers: the queen ensures that at least one of her daughters will have a colony of her own, and at the same time, each worker's reproductive chances increase, because she has half as many other indi- viduals to compete with.

Why can't a worker strike out on her own, build a few cells, store food, and wait for the appropriate time to lay male eggs? Because hymenoptera gather food for their offspring, such small nests would be destroyed by members of neigh- boring colonies, who would steal anything stored in them to feed their own larvae. This danger creates the need to cooperate in guarding and the opportunity for the strong to exploit the weak. Robbery and predation prevent the workers from reproducing outside large colonies ruled by queens.

WHY DO THE WORKERS WORK FOR THE COLONY?

We have seen why workers stay in a colony waiting for an opportunity to reproduce, but that does not explain why they actually *work* for the colony. There are many workers, and most of them are willing to work for the group; why not leave the labor to others?[10] The colony would not be harmed appreciably, and the "lazy" worker would save her strength for the final struggle and the egg laying. One would think that such social parasites would be in a better position in that final struggle and would have more sons than their competitors, and thus such parasitism would spread in the population; yet we do not see that happening. We believe that a worker or a lower-ranking queen enhances her own chances to reproduce by working for the colony.[11]

Altruism and Prestige

We have found a similar relationship in babblers, of course. In babbler groups, a babbler invests in altruism, we believe, because it reliably demonstrates its ability as a rival and its value as a partner, in ways that other members of the group recognize. The gain to the "altruist" is not the benefit that the group derives, but the recognition of other individuals—the prestige he or she wins.

As we saw in the previous chapter, an individual's reproductive chances are determined in large part by prestige. The more a babbler does for its group, the greater its prestige, because its ability to do for the group proves its ability in general. And high prestige—the recognition of its ability by others—spares it the need to fight or even threaten its groupmates when the time comes to compete for the right to reproduce.

The "altruistic" individual thus serves its own interests—it expends effort to demonstrate its quality reliably. Because it does so by engaging in altruism, rather than by pure showing off, as a peacock does with its tail and a bower bird with its bower, the babbler is also showing its interest in continued collaboration with the members of its group. Groupmates pay attention to the dual message because this information enables them to make their own decisions. At the same time, the whole group benefits from the actions that convey the altruist's message.

In both a group of babblers and a colony of social insects, one has to cooperate to survive, and at the same time one competes with others in the group or colony for the right to reproduce. In such situations, anything that allows individuals to demonstrate and assess their own and each other's ability without a physical clash or direct threats, and anything that will help them evaluate a partner's interest in continued collaboration, will increase their chances of success.

Thus there is a ready audience for messages that reliably combine evidence of an individual's ability—as shown by its doing deeds that demand ability—and of its interest in continued collaboration—as demonstrated by its willingness to devote these show-off efforts to its collaborators. By helping, the signaler demonstrates its ability, which is recognized by others; this recognition translates into the rank and prestige that enable it to reproduce more successfully than individuals who receive less recognition. This can become a powerful selective force for behavior that helps the group; the evolution of what looks at first glance like straightforward altruistic behavior is in fact driven purely by the mechanism of individual signal selection.

Queen Pheromones and Prestige

A group of babblers is small enough that the partners can observe each other regularly, and the deeds of each for the benefit of the group are known to all. Things can be the same in small colonies of social hymenoptera:[12] all individuals know each other and are aware of the work done by each, and thus the relative prestige of each. But in a colony of several thousands or tens of thousands of individuals, it is highly unlikely that any individual could observe and remember the deeds of each of the others. If indeed workers do serve a large colony in order to win prestige and acquire rank, there must be a mechanism that demonstrates each individual's ability reliably.

The mechanism that seems likeliest to us is pheromones.[13] We suggest that in a large colony, a worker's ability—her worth—is measured by her capacity to carry queen pheromone. We think that that capacity reflects two factors: the amount the worker's bodily condition allows her to carry, since the pheromone is a harmful chemical; and the amount she manages to acquire, either by force or in return for her services to others. The situation will be more complex and interesting if it turns out, as we think possible, that by working, not only does a worker acquire queen pheromone from the queen or from other workers in return for her services, but she actually increases her capacity to carry queen pheromone as well, since working increases her metabolism.

It is known that the pheromone secreted by the queen motivates workers to serve her. Bees are very eager for the pheromone and lick it off the body of the queen. Workers can get the pheromone directly from the body of the queen when they serve her, or indirectly, from workers who have served the queen. They may also get the pheromone from the larvae they serve, who, it seems, produce similar pheromones.[14] Still, there seems to be some factor that limits the amount of pheromone each worker collects. When the pheromone is dispensed by researchers, all individuals are attracted to it, but only some of them eat it freely, while others eat none or very little.[15]

In some respects, one can liken the queen phero-mone to money in human societies. In a large society, individuals do not know each other personally, but they can tell each other's economic status by the amount of money each has. Only persons of means can hold on to large amounts of money. One carrying money also has to protect it from thieves and robbers—in other words, money testifies not only to its holder's ability to acquire it, either through work or by force, but also to his or her ability to protect it.

The Handicap in Queen Pheromone

It is known that there is a correlation between a queen's ability to lay eggs and her ability to produce pheromone.[16] It is also known that workers kill queens who secrete only a small amount of pheromone.[17] If queens could separate egg-laying from the secretion of pheromone and concentrate when old or in ill-health on the latter to avoid being killed, surely they would do so. Apparently, they cannot—and thus the amount of pheromone they secrete is a reliable indicator of the number of eggs they lay.

According to the Handicap Principle, queen pheromone can be a reliable indicator of egg laying only if overproduction of it—deception—would cause the queen more harm than good. We believe, in other words, that the pheromone comprises intrinsically harmful chemicals. Just as this pheromone advertises reliably the quality of a queen, smaller quantities of the same harmful pheromone can serve as reliable indicators of the abilities of workers as well. Thus a worker's ability to handle pheromone she receives from the queen would be recognized by others and would translate into rank and prestige, which would enable workers to settle their conflicts over reproduction with fewer physical struggles.

It is well known that queen pheromone inhibits the development of workers' ovaries. If the pheromone were just a conventional marker, with no significant inherent properties, one would expect to find an inverse correlation between the quantity of queen pheromone a bee carries and the state of her ovaries. In fact, it turns out that the more highly developed the ovaries of a worker are, the more queen pheromone she carries. Rather than serving as a conventional marker, then, the quantity of queen pheromone carried by an individual seems to indicate her ability to withstand its harmful effects. We believe that the capacity to tolerate queen pheromone, then, varies in individuals according to their physical condition, which is just the way the Handicap Principle would have predicted a chemical signal would work.

Roseler and Honk show that in bumblebee nests, the ovaries of the strongest workers, who are nearest to the queen, are well developed even though their proximity to the queen exposes them to high levels of queen pheromone. The researchers remark that these workers do not hold their high rank *because* of their well-developed ovaries; the workers maintain their rank even if their ovaries are removed surgically. We think that these workers are physically superior to their colleagues, and that this superiority gives them both their high rank and the ability to develop their ovaries in spite of the high concentration of queen pheromone they are exposed to.[18]

In honeybees, the queen pheromone contains unsaturated fatty acids with keto groups. It is known that fatty acids, especially unsaturated fatty acids with keto and alcohol groups, are active, harmful chemicals. They are similar in structure to the female moth pheromones we have already discussed. It may be that queen pheromone is absorbed by the bee's exposed sensory cells and causes a state of intoxication, as we suggested for moths in chapter 9. If we assume that a worker who carries queen pheromone is harmed by it in this or other ways, then the amount a worker can carry is limited not only by her ability to acquire it by force or in exchange for services, and her ability to protect it, but also by her ability to withstand that harm.

Showing off by consuming harmful chemicals like alcohol, tobacco, betel nut, opium, and the like is common among humans. In some societies, men even demonstrate their vigor by drinking naphtha. In his book *The Rise and Fall of the Third Chimpanzee,* Diamond uses the Handicap Principle to explain this phenomenon.[19] Veblen,[20] the sociologist who coined the phrase "conspicuous consumption," likened the men who hang out in bars and pubs, drinking and buying drinks for each other, to American millionaires who show off by funding colleges, hospitals, and museums. One who can drink quantities of alcohol without apparent ill-effect shows reliably his or her good physical condition; one in poor condition would get visibly drunk. Laborers who do hard physical work are notoriously heavy drinkers in many societies.

In sum, we think that a worker serving the colony collects queen pheromone in exchange for her services. Since the capacity of a bee to carry queen pheromone depends on her being in good physical shape, the pheromone is a reliable indicator both of her ability to acquire pheromone and of her current bodily condition. Her tolerance for pheromone is evident to her comrades, and as threats do, it allows her to exhibit her ability and achieve her ends with less actual violence. In fact, possessing the pheromone generally makes even overt threats unnecessary. Moreover, we wonder whether the very act of working for the hive *increases* the worker's ability to carry pheromone, just as, after a year of construction work, a college student may "hold his liquor" better than he did in his fraternity-party days.

The suggestion that queen pheromone is the handicapping signal that proves rank and prestige in a large hive or colony, and that a worker's capacity to carry queen pheromone reflects her quality both as a partner and as a rival, is still a hypothesis. It fits the facts that we know, but the picture is far from complete. Today instruments exist that can measure the quantity of queen pheromone carried by an individual insect. It should therefore be possible to carry out research that tests our hypothesis.

KIN SELECTION THEORY AND ITS DRAWBACKS

Most evolutionary biologists who discuss the social insects take it for granted that workers in a hive or a nest would have a better chance to reproduce if they didn't squander their efforts helping others. In English and other languages, *altruism* is commonly defined simply as devotion to the welfare of others, or regard for others as a principle of action, but evolutionary biologists define it as "a feature that helps another *at a cost to the helper.*" Since the only evolutionary definition of success is successful reproduction, "cost" is defined as a decrease in reproduction. In other words, the evolutionary theories that have been developed so far to explain altruism assume that altruistic behavior decreases reproduction by its very definition.[21]

Now, how can a feature that reduces the reproduction of its bearers evolve, when evolution by its basic nature is the selection and promotion of traits that enable individuals to bring forth more descendants? No explanation of the evolution of altruism makes sense unless it shows that altruistic activity will actually *improve* the reproductive chances of its carriers.[22] In other words, *any* theory that attempts to explain the evolution of altruism must identify some mechanism that compensates the altruist for its supposed loss of reproductive "fitness."

At first glance, group selection seems to explain the altruism of social insects easily: by that theory's logic, if colonies whose workers assist the queen compete successfully with colonies whose workers do not, then the latter will become extinct and only colonies in which workers do in fact work will survive. There is no doubt that the former are more successful; but the question is whether that is enough to support the evolution of such "altruism."

Logically, in a society in which some individuals are inclined to assist, a selfish individual would benefit by being selfish—it would enjoy the prosperity of the colony that arises from the work of others while conserving its own energy for its own reproduction. Eventually, more and more of the group's members will be the descendants of such selfish individuals, and presumably they will be selfish as well, even though this would in the end cause a breakup of the colony. Since like all mutations, those for selfishness—that is, the tendency not to help—appear regularly, one would expect to find groups in different stages of this process, with varying percentages of selfish individuals. But observation shows that this is not the case.

When explanations based on group selection went out of fashion, the question came up again: How could creatures evolve who impair their own reproduction by assisting others to reproduce? This question recurs in every study of social animals, and most dramatically in studies of the social insects. The theory of kin selection was offered by Hamilton as a solution to this problem.[23]

Central to the theory of kin selection is the fact that evolution can be measured by variations in the prevalence of particular genes, which carry traits, in a population. Dawkins in fact went further, seeing evolution as competition between genes rather than between individuals; this is the key assumption in his well-known book *The Selfish Gene*.[24] Kin-selection theory states that one can increase the prevalence of one's own genes in future generations not only by one's own reproduction but also by assisting one's relatives to reproduce, since each specific gene that one carries is likelier to be carried also by one's relatives than by others. Even if, by assisting, one harms one's own reproductive chances, according to the theorists, as long as one improves the chances that the *genes* one carries will appear in the next generation, one is successful.

According to kin selection theory, an individual that invests in helping its sibling to raise two additional children would bequeath to the next generation, on average, the same number of its own genes as it would contribute by raising one child of its own.[25] By this logic, if an individual's help enables a sibling to raise more than double the offspring the individual can have on its own, then that individual is better off aiding that sibling than devoting its efforts to its own reproduction. This kind of calculation, which measures a trait's effect not only on its carrier's reproduction but also on the reproduction of its carrier's relatives, is called "inclusive fitness." Scientists distinguish this from simple "fitness," which takes into account one's descendants only.[26]

Hamilton suggested kin selection as an explanation for the evolution of altruism in animals in general, and he suggested that hymenoptera have some special trait that supports the evolution of social systems by kin selection. After all, true social (eusocial) systems, in which most or all workers are almost sterile, evolved in this group some dozen times independently (several separate times in bees, several separate times in wasps, and in ants); in all other insect groups they evolved only twice, in termites[27] and in some scale insects.

Hamilton suggested that the haplodiploid mechanism of gender determination in hymenoptera plays an important role in the evolution of their social systems. Male hymenoptera have only one set of chromosomes, inherited from the mother, while females have two sets of chromosomes, one of which comprises half their mother's chromosomes and the other all of their father's. The genetic relatedness of a female bee, wasp, or ant to her sister is 75 percent, since they share half their mother's genes and all of their father's. According to Hamilton, it was this genetic relatedness that facilitated the evolution of altruism in hymenoptera: the workers who take care of the young are raising sisters who share 75 percent of their genes, while their own daughters would have shared only 50 percent of their genes.

The theory of kin selection took the scientific world by storm.[28] Researchers of social behavior in animals started measuring the degree of relatedness between individuals. But soon it was discovered that altruistic behavior exists even where there can be no genetic relatedness between the altruist and the one being helped, such as in those species of ants, wasps, and birds in which altruistic social systems link unrelated individuals. What is more, it is known today that often more than one male fertilizes a queen,[29] making for a far lower degree of genetic relatedness among workers and queens in a hive: sister bees who share a mother but not a father have on average a genetic relatedness of only 25 percent.

It was also found that not infrequently, worker bees and ants leave one colony and move to another, where they work as usual, even though the queen and the other workers are not their relatives at all.[30] All this made it necessary to develop another explanation for altruism among unrelated individuals—Trivers's theory of reciprocal altruism, which we discussed in chapter 12. Still, nobody dared challenge the theory of kin selection as the basis of the evolution of altruism. It was hard to give up the magic of a theory that seemed to enable one to calculate mathematically the desirability of altruism.

Parasitism Among Kin, or Haldane's Other Brothers

We believe that the basic premises of the theory of kin selection are as weak as those of group selection. The idea of kin selection was first proposed by Haldane in 1932 in his book *The Causes of Evolution*.[31] If an altruistic trait benefits an individual's descendants and relatives, Haldane wrote, then that trait increases the individual's fitness (defined as the individual's production of successful descendants); the trait thus can spread in the population by natural selection. Haldane offered as an example two brothers walking by a river: one of them falls into the water and starts drowning; the other can jump in and save him. According to Haldane, it would make sense, evolutionarily speaking, for the brother standing on the shore to jump in and save his brother, even though he thus lessens somewhat his own chances to survive and reproduce. The trait to sacrifice oneself for others thus came about through kin selection, according to this theory. (Interestingly enough, Haldane himself said that he had twice in his life jumped into a river to save another person, and that in neither case did he stop to reflect that it did not make sense for him to do so, since the one being saved was highly unlikely to be his relative.)

Haldane's example, however, has the same weakness that the theory of group selection does: it is vulnerable to social parasitism. The basic fallacy in Haldane's argument can be illustrated by a variation on his story about the drowning brother. Let's assume that instead of two brothers walking by the riverside, we have three or four. One of them falls into the river, and another jumps in to save him. The others stand by, doing nothing. All brothers will receive the same genetic gain if the rescue is successful; but the rescuer risks injury or even death, while the ones standing by risk nothing—that is, the slackers' total gain will be greater. The altruist, then, will benefit *less* than a selfish brother, and on average, the number of altruistic genes in the next generation will *decrease* rather than increase.

Motro and Eshel formulated a mathematical model to solve this dilemma, but they did so by adopting an unrealistic premise: that no brother can assume that another brother will jump in the water if he himself refrains from doing so.[32] According to the theory of kin selection, we would predict that each would encourage his brothers to jump in rather than jumping in himself. In reality, we know it is much likelier that all the brothers will jump in to save their drowning sibling, and that each will do his utmost to take on the risk and perform the rescue; no brother will consider that because the others are struggling to do the same, he is risking his genes needlessly.

Thus, group selection theory and kin selection theory have the same flaw: they invite social parasitism. In fact, kin selection is simply group selection among relatives.[33] We find it strange that despite this, many researchers who reject group selection see kin selection as a viable, stable mechanism, like individual selection.[34]

Basing an explanation of altruism on kin selection theory presents two further problems. Research on group-living birds shows that one or two helpers can indeed increase the success of a reproducing couple, but that more helpers than this have no effect on the success of reproduction.[35] By the logic of kin selection, to invest in providing unneeded help is a pure loss both to the helpers and to all their kin. By that same logic, in fact, all the helpers' kin would have benefited if these unnecessary helpers had saved their efforts for their own reproduction or tried to form new groups. Yet the reproducing couple in group-breeding birds often gets assistance from more helpers than they need.

And if the driving force behind the evolution of altruism is the benefit gained by kin, shouldn't the same logic militate strongly against aggression toward kin? Yet we often see violent struggles between kin collaborators. Indeed, these struggles frequently end with one or more of the collaborators being wounded or even killed. The theory of kin selection does not explain why it is that relatives do not avoid harming each other in their struggles.

Gadagkar and Joshi describe a colony of wasps that split into two colonies; before the breakup, the level of aggression in the colony was so high that it caused a sharp decrease in the number of offspring. Only after the breakup did the aggression subside, and the number of offspring then increased a great deal in both new colonies.[36] As Darwin remarked, it is precisely the social insects that are noted for excessive enmity to their closest relatives: mother kills daughter, sister kills sister.[37]

In their monumental book on ants, Hölldobler and Wilson[38] remark that life in an ant colony is a constant struggle for supremacy.

Are Offspring "Kin"?

The debate between supporters of kin selection theory and its critics, like many scientific debates, is not free of emotion and is very hard to conduct with pure logic. In the heat of argument, Dawkins[39] alleged that critics of kin selection refuse to "count" the kinship of offspring and parents in kin-selection calculations; the inclusion of direct offspring, Dawkins insisted, lends plausibility to the theory. In fact, everyone grants the obvious kinship of parents and offspring; but this is beside the point.

The debate over kin selection is a debate over the difference between two measures of an individual's success: straightforward Darwinian natural selection, which measures the success, or "fitness," of an individual by how many of its offspring successfully reproduce; and kin selection, which measures "inclusive fitness" by *adding to* this the individual's effect on the reproductive success of its relatives. There is no point in arguing over what the two theories agree about—that is, the individual's own offspring. The critical issue lies in the *difference* between the theories. The question is whether the reproduction of those among one's relatives who are *not* one's direct descendants has any effect on the evolution of traits.

PARTNERSHIPS AMONG KIN: WHY IT MAKES SENSE TO JOIN THE FAMILY BUSINESS

As we have said, we believe that helpers generally stay in their home group, nest, or colony because that is where their own reproductive chances, slim as they may be, are greatest—and because the more they prove themselves by helping the group, the better those chances become. Why do animals collaborate in particular with their own relatives? Because they know them: the rank order among them has already been established. If for various reasons one's best chance is in a partnership, it often makes sense to band together with individuals one knows well, whose rank relative to one's own is already established.

West describes the formation of a colony of the wasp *Polistes canadensis* by a group of seven queen-sisters.[40] In our opinion, her description shows what each partner gets out of the initial partnership, and what each gains by banding together with her sisters rather than with strangers. Since they were sisters, the social ranking among them had already been established the previous fall, when they were living together in their mother's nest.[41] In the new nest, a division of labor appeared: the top queen stayed in the nest, guarding and taking care of it. The three sisters next to her in rank went out gathering food. The three lowest-ranking sisters stayed by the nest without taking part in any activities. When the top female disappeared, the second took its place and one of the inactive sisters joined the food gatherers.

This arrangement enables each sister to find out where her best chance to succeed lies. The second in rank already knows from experience that she cannot overcome her senior sister. By going out to collect food, she can meet wasps from other nests who are also seeking food. If she finds herself stronger than most of them, there is a good chance she could protect a nest effectively. In that case, it will make sense for her to strike out on her own and establish her own nest.

On the other hand, if the second sister finds herself weaker than other food seekers—each of whom has an even stronger sister guarding the nest it comes from—it will be better for her to go on being a helper and bring food to her home nest. In that case, her best chance will be the possibility that her senior sister will disappear and that she will take her place in an established nest, with helpers and possibly workers. As for the weaker, inactive sisters—as long as they see their stronger sisters coming back to the nest rather than striking out on their own, there is little chance that they, being even weaker, would manage alone; there is no point in trying their luck elsewhere.

West's seven sisters had known each other from the time they hatched and had established an order of rank among themselves before they formed the nest together; they didn't need to spend time and effort and risk injury to establish a hierarchy at the beginning of the season, a critical time. We believe, in fact, that the familiarity and established rank among relatives is the main reason for the prevalence of coalitions among relatives in the animal kingdom.

THE KIN EFFECT

There is no doubt that among social animals, other things being equal, an individual who belongs to a group whose members collaborate reproduces better than one who belongs to a group whose members fight each other. This is true even of collaborations between different species, as in symbiosis. It is also clear that if for some reason it makes sense to collaborate with relatives, then in future generations a scientific tally may sometimes show an increase in the representation of those relatives' genes within the general population. We suggest calling this phenomenon "kin effect."[42]

Kin effect is different from kin selection. In kin selection theory, the *selective factor* that drives the evolution of altruistic traits is the benefit the altruist's relatives gain from the investment that individual makes in them. By contrast, we say that altruistic traits evolve by natural selection; that is, they evolve because in fact they improve an individual's reproductive chances. Investment in the group benefits the investor directly; often the gain to the altruist is increased prestige, which facilitates his or her own reproduction. The kin *effect*—the impact these altruistic traits may have on the reproduction of the individual's kin—cannot be the factor that *selects* the tendency to invest in altruistic acts.

THE PARENTING COUPLE

For animals who reproduce sexually to have offspring, two individuals must collaborate. In chapter 3, we discussed the conflict of interests between the genders during courtship, when each has to choose a partner to mate with and must convince that partner to mate with it. But it is not just the *selection* of a mate that creates conflict between the genders.[1] The pair must then cooperate—mate—in order to bring forth offspring. Some of their roles are innately different: only females can supply eggs, and among mammals they are the only ones who can bear and suckle the young; only males can fertilize. But either or both can take care of the offspring, and that very possibility creates more potential conflicts.

Both mates gain equally from reproducing, but they need not invest equally in the offspring. Often one adult can take care of the young on its own; indeed, each mate stands to benefit if it can exploit the other's interest in the offspring and leave that mate to do the job alone. An individual can increase the number of its successful offspring by choosing the best available mate, by having many sexual partners, or by combining both strategies, which complicates the picture.[2]

In many species, one of the pair leaves the offspring in the care of the other.[3] Usually, the male leaves the young to the care of the female[4] and goes off in search of another female—in other words, he practices polygamy. Sometimes, though more rarely, the female is the one who goes off to find another male and leaves the care of the offspring to the father—or she may mate with several males who all assist her in the care of their common offspring. In such cases, she practices polyandry.

Even when raising offspring takes both partners, their investment need not be the same. True, if one stints and their offspring suffer, both partners lose; but if, by deserting, one of the two mates can form a new or an additional mating partnership, it may well gain more than it loses.

But a deserter does not necessarily gain. In general it has been found that one is more likely to successfully raise a brood with a familiar mate, with whom one has succeeded in raising offspring before, than by starting over with a new mate.[5] Becaused of the pair's familiarity with each other, they are probably spared the effort of checking each other out and coordinating their actions. Also, research on songbirds has shown that young hatched earlier in the season are usually more successful.[6] Thus an individual who deserts its mate and shortchanges their young may well stand to lose considerably more, in terms of its overall reproductive success, than it can gain in another partnership; moreover, there is no guarantee that the deserter will find another partner.[7]

PATERNITY AND MATE-GUARDING

The female usually selects her offspring's fathers. Now that paternity can be determined by genetics, scientists are discovering that many nestlings are not fathered by the males who take care of them. These findings explain the effort males make to guard their female partners during their fertile period.[8] In chapter 12, we described how the dominant male babbler begins following the breeding female closely a few days before she lays eggs; he continues until the day the last egg is fertilized. In the morning, when the female flies to the nest and lays her egg, the dominant male immediately follows her; he sits on the edge of the nest or near it and tries to copulate with her afterward—presumably the best time to fertilize the egg that will be laid the next day. Needless to say, during that time the dominant male tries to prevent any other male, and in particular his second in rank, from approaching the female.[9]

In a thorough study of house sparrows, Moller[10] found that during egg-laying, a female who is not escorted by a male is likely to be attacked by groups of males who try to rape her—a phenomenon that is known in ducks as well. The "husband" tries to guard his mate and chase away any male who approaches the female or the nest. But, as genetic tests show, even close supervision cannot ensure paternity. Moller also reports that many female sparrows copulate willingly with males who are not their mates. The chosen paramour is usually a high-status male with a large black bib, a "status badge" that, as we saw earlier, is a reliable indication of its bearer's quality. Morton[11] found that older pairs of purple martins often let younger pairs settle near them; not infrequently the younger females were seen copulating with their older male neighbors. Zilberman[12] observed sunbirds in Israel copulating with males, usually their neighbors, who were not their mates. Extrapair copulation, as it is termed, turns out to be quite common in nature.

Why does a female copulate with males other than her mate? Many reasons come to mind. There is the risk that males from neighboring territories or solitary males will destroy her nest and kill her offspring, forcing her to lay another round of eggs in the expectation that they will be the ones to fertilize them.[13] To prevent that from happening, the female can try to gain the protection of strong males living near her, who might be willing to shield her offspring in exchange for the chance to father some of them. This phenomenon of killing offspring was first observed with langur monkeys in India.[14] In Israel, Giora Ilani and Yotam Timna found that female leopards mate with several males, and that males kill the young of females who have not copulated with them.[15]

It is also possible that females mate with additional males in case their chosen mates are not fertile: by copulating with another male, a female raises her chances of laying fertile eggs. She might also improve her offspring by copulating with a high-quality male who might not agree to desert his mate for her but is quite willing to father a few additional offspring on the side.

In short, a male who pairs up with a female does not necessarily father all her offspring. A male who wishes to be sure that the offspring his mate bears are his own has to persuade the female of his quality and of his willingness to invest in her and in their common offspring, so that she will not try or agree to copulate with other males.

TAKING CARE OF THE YOUNG TO GAIN PRESTIGE

The need to convince one's partner of one's commitment and quality is so strong that often, rather than exploit each other, the members of the breeding pair compete to provide for their offspring, or even take care of offspring that are not their own. Such competition is evident among herons, egrets, gulls, plovers, and others. For example, when one of the pair is incubating the eggs, and the other comes to replace it, the former may not give up its task willingly; the other has to push it off the nest in order to do its share.

According to Trivers,[16] performing a task that would otherwise be done by another is altruism. Theories of reciprocal altruism, however, as we saw in chapter 12, fail to explain why one of the pair would go so far as to push

the other off the nest. It would seem that this competition over caregiving is meant to demonstrate the quality and motivation of the caregiver and to strengthen the relationship between the parental pair. In fact, when we discuss parasites in chapter 16, we will suggest that by taking care of cuckoo chicks, crows show themselves off as good mates in order to prevent the breakup of the pair.

Competition over the care of offspring, like the competitive altruism of bab-blers, shows both an individual's quality and its commitment to the partnership. Single males often assist widows in raising their offspring for similar reasons; they thus prove their quality and the seriousness of their intent and increase their chances to pair with the widows later on. The widows are desirable mates: they have already proved themselves by having offspring. A single male may assist a pair to raise their offspring in order to increase his chances of mating with the female should her partner die. Sometimes the female's mate is kicked out by the "helper" himself—a phenomenon that has been observed among monogamous species as well as among species that live in groups.[17]

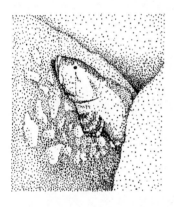

An extreme example of an animal proving its quality by caring for the offspring of others is pro-vided by fish. In the Japanese fish *Alcichthys alci-cornis*, fertilization occurs within the female's body. After the female's eggs have been fertilized by one male, she chooses another and lays her eggs in his nest. Only after she has filled his nest with the first male's offspring does the female agree to copulate with the nest's owner; she then goes off with the eggs he has fertilized and lays them in the nest of yet another male. Most males of this fish thus take care of the offspring of others, as has been proved by DNA analysis of the offspring.[18] Females prefer males in whose nests there are already large egg deposits; thus, the investment of each male in others' offspring increases his attractiveness to females. Indeed, the females of many fish are attracted to males who are taking care of nests that contain plenty of eggs.[19]

OTHER MEANS OF SHOWING OFF TO ONE'S MATE

Showing off to one's mate does not necessarily involve caring for offspring. Black-starts and wheatears nest in small caves and holes in rocks. They pave the entrance to the nest with small stones—often scores and even hundreds of them. Fishman, who studied the white-crowned black wheatear in the Sinai Desert, observed fe-males trying again and again to lift stones too large for them to carry, and in the end making do with the largest stones they could handle.[20]

Moreno and his colleagues studied stone-collecting by black wheatears in Spain.[21] In that species, both males and females collect stones. In the days before egg-laying, they may bring up to three or four pounds (1.5–2 kilograms) of small stones to the nest. The researchers found a clear statistical correlation between the number of stones the pair gathers and the male's participation later in feeding the young, and also between the number of stones and the pair's reproductive success. They checked to see whether the stones themselves were beneficial—whether, for example, they deterred predators or modified the microclimate around the nest. They could not find any such effect, however. They concluded that stone-collecting lets a pair of black wheatears show off their quality to each other and thus strengthens the bond between them.

Leader, who studied blackstarts, found that females collect stones to pave the entrances of their nests. He did not find any correlation between the number of stones collected and the quality of the female and suggested that in the case of blackstarts, the stones may serve a utilitarian purpose—probably as an alarm system that warns of approaching predators.[22] If so, this may well have been the original, utilitarian behavior that in black wheatears evolved into a signal and a test of quality.

DOMINANCE BETWEEN MATES

Cooperation between genders allows each to evolve specific traits. A partnership between parties who complement one another is more effective than one between two identical collaborators. This means, of course, that each of the two has to search for a partner that will complement it. Often the evolution of one trait conflicts with the demands of another: for example, fighting demands a muscular body, while egg-laying demands reserves of fat and protein.

A female duck, who lays some ten eggs in a clutch, spends most of her day during that period straining food out of the water to accumulate the nutrients she needs for her eggs. The way ducks feed, and the large number of eggs that she lays, prevent her from becoming a good fighter: she is unable to protect the territory she and her offspring need, and when she puts her head in the water to collect food, she is exposed to predators. The male, who is not under the same pressure to gather food, can afford to evolve the muscular mass he needs to be able to fight neighboring pairs over territory, to chase off other males who try to copulate with his mate, and to guard her against predators.

The stronger of the two partners in an animal pair—usually the male—has a degree of freedom and of control over shared resources that the weaker partner doesn't. His strength means that he has the ability to establish the best territory he can, then choose the best female that he can find; the weaker gender has to either choose a territory with the mate that comes with it, or choose a mate and settle for whatever territory that mate can hold. The stronger gender can choose a partner, or kick a former partner out of the territory. If he can hold a large and high-quality territory, he may even be able to get an additional female.[23] His greater strength often allows him to dominate the female.

CONDITIONS FOR FEMALE DOMINANCE

In some species, however, the female is stronger and larger than the male, and dominant over him. Female dominance evolves in specific ecological conditions: for example, when on the one hand two adults are needed to take care of the offspring, and on the other hand the "bottleneck" in reproduction is not the female's ability to bear young or lay eggs but occurs at a later stage in the breeding cycle.

The chicks of jacanas and some sandpipers[24] eat different food than adults of these species do. The food adults eat is plentiful, yet even though the females have abundant resources, they cannot use them to produce more eggs. For the food that is available to their *chicks* is *not* plentiful, this is the bottleneck restricting reproduction. In a case like this, where the key constraint on reproduction occurs after the young are hatched, females are freed from devoting most of their resources to egg production. Thus, females can afford to become physically strong and can become bigger and stronger than males.

In birds of prey, too, females are often the dominant partners. Raptors who catch their prey in flight have to be strong and must be quick flyers—which also makes them able fighters. Neither does such a raptor female have to devote all her time to feeding; one pigeon or grouse is enough to satisfy her for an entire day. Thus, she is not adapted to collect food in ways that interfere with her ability to defend a territory. Food for the chicks is a different story. Many raptor chicks die in the nest, and the parents cannot raise more than one or two successfully—an indication that in this case as well, the territory provides more than enough food for the adults, and that it is the scarcity of food for the chicks that limits the number that can be raised.

This explains why a female raptor can afford to become big and strong; but why should the male not be as large, or even larger and stronger? In many species, after all, there is a great deal of overlap between the size of males and that of females. In most species of birds, large females seek out even larger males, who are likely to have better territories, even though they could choose males smaller than themselves and be dominant over them. Among raptors, by contrast, bigger males seek even bigger females, though they could choose smaller females and lord it over them.

A male raptor who selects a female bigger than himself—who is presumably a strong and efficient hunter and has been feeding well—is assuring himself of better offspring. The female contributes more to the offspring than the male does, since she provides not only genes, as he does, but also the egg itself, that is, the environment in which the young develops.[25] Under such conditions, it seems that the higher quality of the female compensates the male for sacrificing dominance and control over their common property, the territory. The female is the dominant not only in raptors, but also in several species of plovers and frigate birds, and in quite a number of fish.

Sometimes, a show of dominance may not be all it seems. Female finches lord it over their mates before egg-laying, while the males are dominant at all other times. It may be that the female's aggressiveness toward the male, which looks like dominance, is actually a test of his commitment to her. The male has to convince the female of his sincere interest in forming a partnership with her; if he does not submit to her wishes, he might find himself raising the offspring of others.

THE PARENTAL COUPLE AS A PARTNERSHIP

The parenting couple is a partnership. As in any other partnership, the choice of a partner and the relationship between the partners depend both on external conditions and on the individuals involved. For example, one may give up dominance in order to gain a better partner and better offspring. And as in any partnership, the two parties have some interests that coincide and others that conflict.

Each member of the partnership—of the breeding couple—can gain from exploiting the other; and indeed, sometimes a breeding partner exploits the other's interest in their common offspring, forcing the other to provide the care that the offspring need. Yet it is precisely because of the parental pair's conflicting interests—because each can take actions that will *not* be in its partner's best interest—that each of them also seeks information on the other's ability and commitment, and provides such information to the other in a reliable manner.

SOCIAL AMEBAS (CELLULAR SLIME MOLDS)

Most scientists who study social systems and communications in multicellular organisms ceased long ago to use group selection as a model—though they still revert to that theory occasionally without realizing it. Microbiologists, by contrast, still routinely use models of group selection to explain the behavior of one-celled organisms such as bacteria and viruses. This approach is justified, they say, because most unicellular organisms reproduce asexually, by division, and live in groups that are genetically uniform; therefore the population counts as a single unit in terms of evolution, the interests of its members are identical, and traits that benefit the group are likely to flourish even if they harm specific individuals.

But this reasoning ignores the fact that the large numbers of one-celled individuals and their rapid reproduction naturally bring forth mutants. Any mutant that can take advantage of others in the population is likely to increase more rapidly than those other cells. We say, therefore, that microorganisms too have to be studied from the point of view of individual selection.

THE LIFE CYCLE OF CELLULAR SLIME MOLDS

Social amebas, or cellular slime molds, of the genus *Dictyostelium* live in the ground as unicellular organisms, each on its own. But when food is scarce or conditions are otherwise stressful, they gather together into aggregates of hundreds or even tens of thousands of individuals. A few hours later the aggregate starts migrating; at that stage, this "slime mold," which can reach the length of one millimeter, looks to the naked eye like a small, slimy slug. In a petri dish in the laboratory, where most experiments on slime molds take place, this migration can last for several hours; when it stops, the aggregate gathers together again and the "slug" changes into a fruiting body—a stalk with a ball of spores on its tip.[1]

During this process, most of the amebas at the front of the "slug"—not quite 20 percent of the total—differentiate into a prestalk type; then they form the stalk and die. The amebas farther back—not quite 80 percent of the total "slug"—differentiate into a prespore type, climb the stalk and become spores; and the amebas in a last small group, which up to now were the very end of the "tail," also change into the prestalk type; these become the basal disk at the foot of the stalk and die. The process by which the population is differentiated into two types of cells is considered a model for the study of differentiation in primitive multicellular organisms and is studied as such in many laboratories around the world. Detailed information on the process, and on the chemical signals facilitating communication among the amebas and coordinating their behavior, has been collected in these studies.

FORMING THE STALK: ALTRUISTIC SUICIDE?

The amebas that die in forming the stalk are considered altruistic, since by dying they help the rest of the population to survive. The stalk seems to improve the chance that the ball of spores will manage to hitch a ride on a passing insect and thus reach a new supply of food; it may also be that the stalk protects the ball of spores from predators in the soil, and that any single spore is likelier to survive in the ball than it would on its own in the soil.

But why do some of the amebas "sacrifice" themselves to form the stalk? This does not present a problem for someone who believes in group selection; but as we have already stated, group selection models are inherently unstable. If the mechanism that causes the amebas to form the stalk evolved because the stalk benefits the spores, then what happens when a chance mutation leads even one ameba to disregard the chemical command to change into the prestalk type? That ameba will change into a spore in defiance of the command; it may survive, unlike the others, and such defiance will spread in the population. But we don't find such a trait spreading in any population that has been studied.

But if we try to explain the phenomenon of stalk creation in terms of individual selection, we are faced with another problem: Why do the amebas that end up in the stalk develop features that eventually kill them? Why don't they instead try to get a new lease on life by becoming spores? This question is all the more interesting if we note—as experiments show—that the process by which particular amebas become stalk can be arrested and the fate of those amebas altered. When a researcher cuts one slime-mold "slug" in two across its width, each piece on its own usually produces a small fruiting body, which has the same form as a regular fruiting body.

When the slime mold is cut in two, most of the amebas that were in the front of the original slug, which would have formed its stalk and died, become spores instead. On the other hand, 20 percent of the other cut piece—amebas that were in the middle and back of the original slug, which would all have developed into spores in that slug—end up becoming stalk, even if they show signs that they were in the process of turning into spores when the original slug was cut.

The destiny of individual amebas, then, is not sealed as soon as the slime mold forms—so again, why do almost one fifth of those amebas give their lives to form a stalk that will help their fellows survive? Can individual selection account for the evolution of the fruiting bodies of social amebas? This question led us to conduct research on the matter with Daniella Atzmony of Tel Aviv University and Vidyanand Nanjundiah of the National Research Institute of Bangalore, India.[2]

THE INDIVIDUAL SELECTION HYPOTHESIS

We have reached two related hypotheses. First, that these stalk-forming amebas are adopting a strategy that under some conditions, however limited, actually increases their chances of survival. Second, that if these amebas had attempted to become spores, their chances of survival would in fact have been even smaller. This is essentially the same explanation we offer for "altruism" among babblers and in the social insects: the assistance offered by nonreproducing individuals to reproducing individuals actually increases the former's chances of reproducing.

The individual amebas gather together when food or water is scarce, and their chances of surviving on their own in that place and at that time are nil or close to it. Aggregates—slugs—have a better chance to survive: they can move quickly, bridge gaps between soil particles, and reach a new source of food, or climb up to the surface of the soil, where their spores will have a greater chance to reach a better environment. Experiments show that the probability that a particular ameba will become a spore depends on the amount of nutrients it has stored compared with others in the same aggregate: the ones richer in stored nutrients become spores. Amebas poor in nutrients—for example, ones that have just divided, whose resources are therefore depleted—adopt the prestalk type. This effect is very pronounced when one mixes populations of amebas with artificially altered amounts of stored nutrients: amebas raised on sugars, and thus rich in nutrients, are much more likely to become spores than amebas raised on bacteria, which are poorer in nutrients—even though almost 80 percent of each population would have become spores if each had been allowed to form a fruiting body on its own.

We believe, then, that amebas forming a slime mold differentiate by the following process. When amebas gather into an aggregate, there is no way to tell in advance how each will compare with others. As a result, they all behave in the same way: they broadcast their interest in getting together by waves of a chemical called cAMP,[3] and they all home in on the principal source of those waves. Once they get together, it is apparently too late for any of them to change course. They cannot leave and seek another aggregate, and it seems that at that point, their chances on their own are nonexistent. Once amebas enter an aggregate, their best chances are in that aggregate—even for those who turn out to be less well-nourished than others; the odds in the aggregate might be against them, but the odds outside it would be worse still.

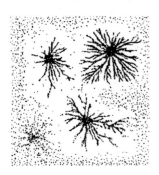

If the only way to survive when food is scarce were to become a spore in a fruiting body, it would be pointless for any individual ameba to adopt the prestalk strategy. However, nutrient-poor amebas have other chances, however unlikely. By joining a migrating slug, they increase their odds of survival, since the "slug" provides some protection against drying out, and since it moves more quickly than an ameba can move on its own. One way to survive is to become a spore, and indeed, as we predicted, it has recently been found—after we formed our hypothesis—that a few of the amebas that start out as prestalk actually end up becoming spores.[4]

One can imagine other lucky events that would enable a prestalk ameba to survive. The migrating slug may meet with another slug that contains even weaker amebas. When that happens, the weakest amebas in the first slug, which were headed toward stalk formation, may well end up among the almost 80 percent of the *combined* slug that are stronger than the rest and become spores. Moreover, the amebas that do not become spores do not necessarily die right away. As long as they are alive they have some chance to survive if the slug quickly gets to an environment rich in food. For stalk-forming amebas, these various chances of survival, however slim, are better than the odds would have been had they tried from the outset to become spores in the presence of their stronger peers. Obviously, this subject requires further research not only in the artificial environment of petri dishes in the lab but in the field.

DIF AS A POISON

The differentiation of amebas into the prestalk type follows the emission of DIF, a chemical produced by the prespore amebas. And it is here that our hypothesis that the slime mold is a product of individual rather than group selection becomes as well a question about the function of DIF as a signal.

Most scientists consider DIF to be a "signal," which "tells" some of the amebas that they have to develop into the prestalk type. Group-selection models have no difficulty with this: DIF causes some amebas to differentiate into prestalk amebas, they say, and these prestalk amebas end up assisting the prespore amebas for the good of the group. But if one assumes the process evolved by individual selection, then two questions come up. First, why do the prestalk amebas "obey" the signal and differentiate into the prestalk form, a process likely to end with their death? Second, why do *all* prespore amebas bother to emit DIF? After all, if the amebas around them are already emitting DIF, a single ameba would still reap the benefit of that emission even if it did not emit any itself—and it would save itself the cost of producing and emitting the DIF. How does each *individual* prespore ameba benefit by emitting DIF?

DIF eventually causes the prestalk amebas to die; for that reason, we suspected that it was a harmful chemical—a poison. This prediction turned out to be true.[5] We believe that DIF's poisonous properties are its very reason for being. We think that the DIF is a poison used by spores to avoid predation, and that the prespore amebas manufacture it for their own individual benefit—probably to protect themselves.

Chemicals that are secreted to keep others away—allelopathic chemicals—including antibiotics,[6] are usually understood to be a tool that enables a population resistant to those chemicals to get rid of populations of bacteria that are sensitive to them. This model, again, is based on group selection: any given individual could save itself the trouble of manufacturing the chemical and rely instead on the secretions of others in the population.

We, on the other hand, ask rather how the individual cell benefits by secreting an antibiotic or a toxin. Perhaps the antibiotic helps the individual cell protect its living space and its immediate surroundings against other individuals that might prey on it or cause it other harm. If so, then although antibiotics may help an entire population against another, this is not what drives the evolution of antibiotics but is rather a side effect. In fact, in natural populations, the amount of antibiotics secreted by cells is minute and cannot kill off other populations. Cells that manufacture large amounts of antibiotics[6] are produced only by means of a strenuous selection process conducted by researchers in the lab.

It is not uncommon for microorganisms to use poisons to protect dormant spores, which would otherwise be easy prey. The effort needed to produce the poisons, and the danger and inconvenience the poisons themselves pose to the pre-spore cells that produce them, are compensated for by the protection the poisons provide. We suspect that this is the primary reason for the manufacture and emission of DIF; the effect on lower-quality amebas—the ones with fewer stored nutrients—is secondary. If this hypothesis turns out to be true, then DIF is not a signal but rather a straightforward tool. For example, rain causes people to take cover, but nobody would suggest that rain is a *signal* that appears *in order to* cause people to take cover.

Since we suggested this hypothesis, Atzmony and Nanjundiah have found indications through bioassay that DIF is indeed present in spores.[7]

If high concentrations of DIF kill amebas, lower concentrations are probably harmful as well. The use of DIF forces amebas to devote resources to surviving its influence and breaking it down. In weaker amebas, we believe, these efforts create the prestalk phenotype, which increases these amebas' ability to survive the poison's influence in the short term; we have suggested that by trying to form a spore, a weaker ameba actually reduces its chances of survival, and it is the presence of DIF more than anything else that creates this stark reality. Collaboration is crucial to all the amebas—if they don't form the slug none will survive—and that is what makes it possible for the stronger ones, those that can withstand the poison and form spores, to exploit the weaker ones.

Cellular slime molds are not unique. Myxobacteria, too, form communal fruiting bodies for their spores in reaction to hunger.[8] In the process of forming the fruiting body, some 80 percent of the cells are killed by fatty acids emitted by the bacteria themselves; apparently the stronger bacteria emit the fatty acids and kill the weaker in the process of becoming spores. In this case too, it seems to us, it is because collaboration is necessary to all that stronger individuals are able to exploit weaker ones; again, to understand the process, we have to understand how each spore-forming bacterium cell benefits by emitting the special fatty acids that kill its comrades.[9]

THE DIFFERENCE BETWEEN PRESTALK AND PRESPORE AMEBAS

What causes the different reaction to DIF in prespore and prestalk amebas? As we said, the difference between the two is not in genotype, but rather in phenotype: the prestalk amebas are relatively short of stored nutrients. We think that in order to breathe, they have to break down proteins from their own bodies, unlike their prespore siblings, which use their stores of glycogen. Could it be that the DIF interferes somehow with the breakdown of proteins? We started exploring this hypothesis with the aid of Dorit Arad, a chemist who studies the relationship between a given molecule's structure and its activity.

But while we were doing so, Shaulski and Loomis[10] found that DIF impairs the activity of mitochondria—the organelles, or intracellular organs, that produce energy by breathing. They also found that in the process of differentiating into the prespore type, prespore amebas produce additional mitochondria. Perhaps the new mitochondria develop to supplement those weakened by DIF. If so, it may be that the DIF prevents the weaker amebas from becoming spores because they do not have the resources to build new mitochondria. It may still be that DIF also interferes with the breakdown of proteins; we are still awaiting conclusive evidence on the matter. In either case, DIF has indeed been shown to be a poison.

If DIF were simply an arbitrary, conventional signal, then a mutation that made an individual ameba disregard it would be beneficial to that ameba. But if, as we think, DIF is a poison that actually prevents weaker amebas from turning into spores, then a weak ameba cannot under any circumstances afford to try and form a spore in its presence. It would make sense to those weaker amebas to go the prestalk route and survive as best they can, for as long as they can. In fact, some of the specific properties of prestalk amebas seem to have evolved precisely to enable these cells to withstand the poisonous effect of DIF: they emit an enzyme that breaks down the DIF, and they move away from the prespore amebas that emit DIF and into the front end of the slug, where the DIF concentration is lowest. The emission of DIF, then, provides information to the weaker amebas; but DIF emission did not *evolve* in order to do this, but rather to enhance each stronger ameba's chances of survival.

SOME REMAINING QUESTIONS

If DIF is a poison, that would explain why prestalk amebas attempt to break it down and move away from it. But it does not explain the considerable effort prestalk amebas devote to moving the slug and building the stalk—the behaviors that have traditionally been called "altruistic." The fast movement to the surface of the soil benefits the prestalk amebas overall, but that in itself does not explain the effort that each individual ameba devotes to the movement. Our model will not be complete until we can explain these aspects of the social amebas' life cycle.

The movement of the slug involves the emission of cAMP, which attracts amebas. When the amebas first gather together, each ameba reacts to cAMP by emitting cAMP itself, and thus the large aggregates of amebas are formed. But with the formation of the slug, it seems that the prespore amebas stop emitting cAMP, while the prestalk amebas at the front tip of the slug continue emitting it and thus get the rest of the amebas to follow them. To understand the movement of the aggregate or slime mold, one has to understand not just the effect of cAMP on the amebas that receive it and respond to it, but more important, how each individual ameba that emits cAMP benefits by doing so. Unfortunately, we don't have answers to these questions yet, but we hope that by posing them we will have made a useful contribution.

WHEN IS A CHEMICAL A SIGNAL?

Until now, the approach of most scientists has been to explain the value of each chemical "signal" according to its effect on the individuals reacting to it, rather than the benefit it brings to the individual emitting it, or the reliable information it provides about that individual. This has led, in our opinion, to a lumping together of chemical *signals*—chemicals that evolved to facilitate communication between individuals or between body cells—and chemicals that evolved for straightforward utilitarian purposes, whose effect on other individuals is either incidental or coercive. In order to understand chemical signaling, one must differentiate between these categories.

We believe that chemical *signals* conform to the rules laid out in earlier chapters for all other signals: the signal must benefit both the sender and the receiver; the signal must be reliable—which means it must impose a real cost on the signaler, a cost too high for a cheater to pay; and the specific cost of the signal must relate logically to the content of the message it conveys. Investigations of chemical signals among one-celled organisms, and chemical signals in general, carried out with these principles in mind may lead us to strikingly new understandings of how such signals work.

CHAPTER 16

PARASITE AND HOST

AN ARMS RACE OR A STATE OF EQUILIBRIUM?

When a parasite attaches itself to a new host species, both species rapidly develop new traits. The parasite tries to exploit the host as much as it can, and the host tries to protect itself against the parasite. Dawkins and Krebs compared this relationship to an arms race between two superpowers, with each manufacturing new and more sophisticated weapons to counter those the other develops,[1] and it is common for scientists nowadays to portray hosts and parasites as being in the midst of such an arms race.

We see things differently. There are undoubtedly "arms races," but we think they quickly reach a stalemate. We begin by assuming that unless there is a specific reason to think otherwise, most host–parasite systems that exist today are in a state of equilibrium: however hard it may have tried, the host has not managed to get rid of the parasite, and obviously the parasite has not exploited the host to extinction. To understand this balance in each particular case, one needs a thorough, detailed knowledge of the lives of both host and parasite. What is it that stops the host from evolving the ability to prevent the parasite from exploiting it? What is it that prevent the parasite from exploiting each host to the point of zero reproduction, or from exploiting more than a certain percentage of its potential hosts?

EUROPEAN CUCKOOS AND REED WARBLERS

Let's take the European cuckoo and the reed warbler as an example. The cuckoo lays eggs in the nests of reed warblers. Immediately after hatching, the cuckoo nestling pushes its host's eggs or nestlings out of the nest; thus, a nest parasitized by a cuckoo does not produce even one warbler. In response, reed warblers evolved the ability to identify the cuckoo's egg and push it out. This ability to recognize and reject cuckoo eggs evolved, it seems, in direct reaction to the cuckoos' parasitism: in Australia, where there are no European cuckoos, none of the reed warblers rejected cuckoo eggs placed experimentally in their nests.

Once the warblers evolved their new method of self-defense, the European cuckoos in Europe and Asia, in their turn, evolved eggs similar in size and color to the eggs of reed warblers. European cuckoos parasitize other songbirds as well and are divided into different subgroups (genets) according to the host they parasitize. In most cases, the color, size, and shape of the cuckoo's eggs are similar to those of the host species.

In areas inhabited by cuckoos, some species of songbirds reject all cuckoo eggs, and cuckoos thus cannot parasitize them. Not so with reed warblers: in almost all areas inhabited by both reed warblers and cuckoos, some reed warblers reject cuckoo eggs and others do not. Davies and Brook[2] assumed that this current situation is a temporary stage in a long evolutionary process—an "arms race"—at the end of which all reed warblers will reject cuckoo eggs and the cuckoos will not be able to parasitize reed warblers at all.

Is this really the case? As we said in the discussion of prey and predators in the first chapter of this book, anything can be explained away as not yet having evolved to its logical conclusion. Yet all reed warbler populations, both in Britain and in Japan—and, presumably, anywhere in between—include some individuals who reject cuckoo eggs and others who accept them. Thus, the "mutation" to recognize cuckoo eggs and reject them has apparently already appeared in all reed warbler populations. Since the benefit of rejecting cuckoo eggs is so great, why has this "mutant" not yet spread throughout the population? Why is it that all over Europe and Asia, from Britain to Japan, we do not know of even one population of reed warblers all of whose members reject cuckoo eggs? What is it that prevents some reed warblers from doing so?

Lotem, who studied cuckoos and reed warblers in Japan with Hiroshi Nakamura and with us, suggested that the difference between reed warblers who reject cuckoo eggs and those who accept them is not a genetic trait in the former that has "not yet" spread throughout the population; rather, it is a phenotypic difference—a difference in age. Most reed warblers who reach their breeding area early reject cuckoo eggs. All these birds have new tail feathers, which indicates that they molted the previous fall and thus that they are two years old or older. By contrast, reed warblers who arrive at their breeding area later, most of whom do not reject cuckoo eggs, have very worn tail feathers. Both the state of their tails and their late arrival indicate that they are young, and that this is their first year of nesting.

What difference does age make? The eggs of reed warblers, like the eggs of many bird species, vary; each female lays eggs that have their own distinct color, pattern, and shape. Some are darker, others lighter; some have larger spots, others fine dots. Cuckoo eggs may look like reed warbler eggs generally, but they cannot look exactly like the eggs of a specific female reed warbler.[3]

Lotem suggests that a reed warbler who has nested before has learned to recognize her own eggs and so can safely reject cuckoo eggs; inexperienced reed warblers, on the other hand, cannot. If they were to reject any egg that looked suspicious, they might inadvertently destroy their own eggs—which is apparently a greater risk than the 5- to-20-percent chance that they will fall victim to a cuckoo in that first nesting. Lotem conducted experiments that showed that indeed reed warblers learn to recognize their own eggs. The relationship between reed warblers and cuckoos is thus not an ongoing arms race, in which some warblers are at a disadvantage because they "haven't yet" acquired a genetic trait enabling them to get rid of cuckoo eggs; rather, they have reached an equilibrium in which each individual does the best it can to succeed in reproduction.

Of course, one still has to explain why reed warblers who don't destroy cuckoo eggs take care of the cuckoo's nestling after it is hatched.[4] The parent warbler does this both in the nest and out of it, for some four to five weeks, by which time the young cuckoo's size, shape, color, and calls are all very different from those of reed warbler nestlings.

GREAT SPOTTED CUCKOOS AND CROWS

Crows, unlike reed warblers, never try to reject the eggs or nestlings of the great spotted cuckoos that parasitize them.[5] Crows lose only a part of their reproductive potential in a parasitized nest: the laying cuckoo usually breaks or ejects only one of their eggs when she lays her own and leaves the rest alone. For its part, the nestling of the great spotted cuckoo does not push the crow's young out of the nest, but it does compete with them for food and thus causes them considerable harm. Yoram Shpirer, who studied cuckoos and crows in Israel, found that an average unparasitized crows' nest fledges two successful offspring, while on average a nest that contains a cuckoo nestling fledges only one successful crow.

Crows have a reputation for being among the most intelligent of birds. They distinguish between different foods as well as between different people, and they recognize enemies. One would think that crows could increase their breeding success a great deal by learning to discriminate between their own eggs and cuckoo eggs and getting rid of the latter. But they don't do this. A common explanation is that cuckoos have just started parasitizing crows, and the crows "haven't yet" evolved the ability to fight against them. But we think this is unlikely. Great spotted cuckoos parasitize various populations and species of the crow family (Corvids) all over Asia and North Africa, and throughout this large region no population shows any resistance to them, with one exception: magpies in Spain. Many magpies in Spain reject eggs of the great spotted cuckoo, but not all of them do.[6] We can think of two alternative models that can explain the constraints that prevent crows from resisting cuckoos: the prestige model and the Mafia[7] model.

THE PRESTIGE MODEL

The prestige model suggests that the host devotes effort to rearing the parasite's offspring to gain prestige in the eyes of its own partner and to practice working in tandem with that partner. Taking care of the offspring of others to improve one's own chances at reproduction is not so unusual;[8] we discussed more than a few such cases in chapter 14. If an individual can improve its reproductive chances by raising the offspring of others of its own species, it makes sense that individuals might also gain by raising parasites.

It is well known that pairs who have already successfully raised offspring together are likelier to succeed than those who have not; this is true partly because pairs who do not succeed in raising offspring tend to break up.[9] It may well be that while a new partnership offers better prospects than one that has failed, it represents more of a gamble than does sticking with an old partner with whom one has already established a successful working relationship.

The research of Yom-Tov and others on crows[10] suggests that the reproductive success of crows depends largely on their ability to get hold of a good territory and keep it. We assume that a well-coordinated pair of partners is much likelier

to achieve this. What, then, are the options for a pair of crows in whose nest a cuckoo egg has been substituted for one of their own? Once that happens, they cannot raise as many young crows in that nest as they otherwise would have, for one of their own eggs has already been destroyed. And because they have been less than fully successful in raising their young, they also run the risk of splitting up. Yet by devoting care to the cuckoo nestling, they may be able to convince each other of their quality as parents and thus stay together. If they do not break up, they stand a good chance of having a fully successful nest next time. If they "divorce," on the other hand, each partner would have to find a new mate and might fail altogether; at best, each would be likely to have difficulty raising offspring next year as well.

Fledgling cuckoos that come out of crows' nests behave as though they know that the only reason their hosts take care of them is to gain prestige. They beg for food very loudly and pursue their host–parents far more aggressively than their crow nestmates. Rather than try to imitate the crow nestlings' behavior, the young cuckoos stand out by noisily importuning. We believe that the crows' real offspring don't need to beg as much, since their parents derive a direct reproductive benefit from feeding them. But the crow does not gain the same way from feeding the cuckoo, of course; the only gain it gets from feeding that "pest" is proving to other crows—including its mate—how good a provider it is. Thus, the cuckoo nestling has to draw others' attention to the act of parenting in order to get fed. Of course, this noisy behavior is dangerous: the young cuckoo is more likely to be noticed by predators. This risk is the price it has to pay for its food.

Needless to say, we would not dream of asserting that the cuckoo and the crows are aware of and consciously calculate all these factors of risk and gain. Rather, fledglings who strike the best balance between risky begging and making do without begging are more likely to grow up, become successful adults, and have offspring who will spread their traits in the population. Similarly, adult crows who invest the right amount in convincing others of their quality as providers are the ones more likely to find and keep better mates and to have more successful off-spring. Ultimately, as with any trait, cost and gain must be balanced—and young cuckoos have to invest more in begging than young crows do, in order to get sufficient care out of their host–parents.

THE MAFIA MODEL

The Mafia model is very different from both the "arms race" model and the pres-tige model. In this case, the parasite is not *tricking* its host but rather *forcing* the host to take care of its offspring. It is well known that cuckoos prey on the eggs and nestlings of other birds;[11] thus, it would seem they could enforce their will as effectively as any mafioso. Suppose a cuckoo were to revisit nests it has laid eggs in, leaving alone the ones where it finds its nestling living but eating the entire contents of nests where its offspring has been rejected. In that case, the discrimi-nating host will come out the loser, and the ability or propensity to discriminate will not spread in the population.[12]

What would a cuckoo gain from "punishing" hosts that recognize and reject her eggs? She cannot be doing so in order to help her rejected offspring, which is already dead. If we cannot find some other direct benefit to the individual cuckoo who does the punishing, then the Mafia model will be based on group-selection arguments—that is, it would be good for cuckoos as a group but would demand that the individual make an effort from which she herself does not gain. We have already explained why we cannot accept models based on group-selection logic. In this case, though, several direct benefits to the cuckoo who does the punishing come to mind. Perhaps the cuckoo is forcing the hosts to lay again, which will give her another opportunity to lay an egg in their nest. Or perhaps she just gains a good meal.

Lotem did not find evidence of such behavior among European cuckoos in Japan; neither they nor the great crested cuckoo that parasitizes crows in Israel seem to punish unsatisfactory hosts according to the Mafia model. But Soler's research in Spain seems to point in precisely that direction: in Spain, great crested cuckoos return to the nests of magpies, and if their eggs have been rejected they harm the magpie nestlings,[13] presumably to teach magpies that nest in their territory not to reject cuckoo's eggs. This fascinating phenomenon is now under study by Soler and his colleagues.

The relationship between the parasite bee *Nomada marshamella* and its host, the bee *Andrena sabulosa*,[14] illustrates the logical next step in the Mafia model. The parasite bee enters the nest of its host without resistance from the owner and lays her egg at the side of the host's pupa. It is highly probable that the host recognizes the parasite's egg as such, but she continues bringing food to the nest, takes care of her own offspring, and does not try to get rid of the stranger.

Why doesn't the host get rid of the parasite's egg? Apparently the parasite, who doesn't need to collect food, remains near the nest almost all the time. Once she has laid her own eggs in it, it is worth her while to protect the nest against robbers and other parasites. From time to time, she threatens the host with the stinger at the end of her abdomen, which can fold to the front like a scorpion's. The parasite, who doesn't need to carry food, can afford to evolve an abdomen flexible enough to be an effective fighting tool inside the tunnels of the nest; the

host, who has to carry pollen and nectar from plants in the field to the nest, does not have such an abdomen. Since the parasite can fight better than the nest's owner, she is able to prevent others of her species, and individuals of other species, from robbing the nest. The host accepts the damage done by this one parasite and acquires in exchange a partner who protects her against others.[15] This is supposed to be the relationship of the Mafia to its clients, and this was often the relationship between knights and their serfs in the Middle Ages: the knights levied taxes, often very heavy ones, and demanded service, but they protected the serfs against the even greater damage that other knights and robbers might have inflicted.

ACCEPTING A PARASITE TO MINIMIZE DAMAGE

Many parasitic birds peck or push out of the nest only some of the host's eggs when they lay their own. Once this significant but limited damage is done, getting rid of the parasite's egg will not repair the damage. Offspring hatched earlier in the season are usually more successful, and it is preferable to continue taking care of the nest and to make do with fewer, more promising offspring than to desert the nest and attempt to build a new one—which itself might fall prey to the parasite.

Sorensen describes such a relationship between two species of ducks, the canvasback (*Aythya valisinaria*) and the redhead (*Aythya americana*), a parasite that lays in the former's nests.[16] The host recognizes the parasite as such, and when the parasite approaches her nest she treats her as an enemy and attacks her. But once the parasite settles on the nest, the host stops resisting and the two females end up laying eggs together. The result is a mixed nest, with no more ducklings on average than a normal brood. The host incubates the combined eggs alone and seems to have no trouble looking after all the ducklings and leading them to food sources. According to Sorensen, fighting on top of the nest would have caused more damage to the host's eggs than the presence of a few parasite eggs does.

Song sparrows are parasitized by brown-headed cowbirds; older song sparrows seem actively to help the cowbird find their nests by mobbing the parasite when it approaches. As a result, female song sparrows two years old or more are parasitized more heavily by cowbirds than are younger song sparrows, who do not mob the cowbirds. Yet older song sparrows are more experienced than younger ones and on average reproduce more successfully. Obviously, it is to the parasites' advantage to know who these better surrogate mothers are—but why do the older sparrows help the cowbird find their nests? Smith and his colleagues, who described the phenomenon, suggested that the system is not yet in balance, and that eventually evolution will lead song sparrows to stop mobbing cowbirds.[17]

We suggest a different explanation: when a cowbird lays her egg in the nest of a song sparrow who is herself still laying eggs, the cowbird does not destroy all of her host's eggs. But when she finds eggs in the later stages of incubation in a sparrow's nest, then the cowbird tends to peck more or even all of them—apparently in order to make the host lay again, so that the cowbird's eggs will be incubated from the day the sparrow starts incubating her own. Older, experienced song sparrows start nesting earlier in the season than younger ones. If we assume that eventually the cowbird will find most of the song sparrow nests in her territory, including replacement nests a sparrow may build after her first nest fails, then it makes sense for song sparrows to expose their nests to her early, so that the parasite will lay in their nests rather than destroy all their eggs.

Thus it seems that older song sparrows who mob the cowbird and thereby reveal their nests' location would rather accept a certain amount of damage than pay a heavier price for attempting to evade or reject the parasite. If the damage caused by the one foreign nestling is not too great, and if it is better for song sparrows to fledge offspring early in the season rather than later, as is the case with many birds, then it seems that song sparrows who mob cowbirds are using the best strategy available to them when cowbirds are present, as in western Canada today. Why, then, don't younger female sparrows mob cowbirds? It may be that they are not able to raise both their own young and the parasite's nestling. Their only recourse is to lie low and hope the cowbird will miss them.

In our opinion, the coexistence of parasites and hosts often depends on the fact that a host stands to lose more in an all-out struggle with the parasite than it loses by tolerating the current level of parasitism. In each case the result is a dynamic balance, and any change in conditions—any weighting of the scales on one side or the other—might jar it out of equilibrium.

NEUTERING THE HOST

Many parasitic worms and fungi attack the host's reproductive organs and make it sterile or nearly so.[18] Then even if the host later manages to get rid of the parasite, the host cannot reproduce and pass on to its descendants the traits that enabled it to free itself. We believe that this strategy of sterilizing the host can evolve only if the *individual* parasite gains directly from it; again, if the species as a group benefits rather than the individual, we are left with a model based on group selection. But the individual parasite stands to gain plenty in this case: because the host is prevented from devoting itself to its own reproduction, more of its resources are available to the parasite. In this case, the only defense a host species can pass on to future generations—the only defense that can spread among the host population by evolution—is strategies that prevent the parasite from gaining even the first foothold, or that prevent it from sterilizing the host.

FROM PARASITE TO COLLABORATOR

Schwammberger[19] studied the relationship between the parasite wasp *Sulcopolistes artimandibularis* and its host, the wasp *Polistes biglumis*. This parasite, it turns out, takes over a main nest, where she lays eggs; she then takes over neighboring nests, where she does not lay eggs but rather steals large larvae and pupae, which she brings to the main nest as food for her own and her host's offspring. The parasite protects her main nest against other robbers, a very real benefit to the host. Schwammberger found that the presence of this parasite was beneficial to the host. True, when at the end of the season he compared the success of nonparasitized colonies with those that had this parasite, the former produced more offspring. But when he took into account all the colonies that failed during the season, he found that the chances that a nest would survive were much higher if it had such a parasitic collaborator. It turns out that the parasite provides crucial help early in the season, when the host does not yet have daughters to share the load: by protecting the nest, the parasite enables the host to undertake longer hunting trips. In fact, in this case the parasite fulfills the function same-species cofounders do in other species.

Sometimes a host can choose among parasites and pick the one least likely to damage it; that one might even protect it against the others. As we saw above, this is a logical outcome of the Mafia model and can lead to symbiosis between species, like the partnership between humans and dogs. When in distant prehistoric days humans started bringing food to their camps and storing it, rodents and insects started gathering there, and in their wake predators that lived both on the humans' leftovers and on the rodents and insects they attracted. Among these larger predators was the wolf, who, because of its preexisting social traits, was easily integrated into human society.

The human-allied wolf refrained from preying on weaker humans such as young children and chased other predators, including other wolves, away from the territory it shared with the human beings. In exchange, the humans shared their food with the wolf-dogs. A dog fed by humans has an advantage over its wild kin: it is protected and fed and thus can afford to become larger and to take risks in fights that it could not afford to take if it had to find food or catch prey on its own. By cooperating with humans and feeding at its owner's table, the dog gave up some of its ability to live independently.

THE LESS VIRULENT PARASITE AS A COLLABORATOR AGAINST ITS VIRULENT VARIANT

The partnership between a host and a nonvirulent parasite is like that between human and dog. The support provided by the host lets the nonvirulent parasite evolve traits that enable it to compete successfully with virulent variants of its own species. It makes sense, then, for the host to invest in the nonvirulent parasite and support it against the virulent parasite.

Sometimes a parasite can demand support by holding the threat of virulence over its host's head. The bacterium that causes diphtheria exists in the throats of most humans in a nonvirulent form. A toxin gene resides in plasmids inside the bacterium,[20] which can manufacture a killing toxin; but in most cases a special repressor protein, manufactured by a gene in the bacterium's chromosome, keeps the bacterium's toxin gene inactive.

To manufacture the repressor protein, the bacterium needs iron that it gets from its human host. When the bacterium does not get enough iron, it stops manufacturing the repressor protein, and the plasmids start producing the diphtheria toxin, causing the deadly disease. As long as the bacterium receives the nutrients it needs—including iron—it does not produce the toxin, so it is worthwhile for a human body that can spare them to discharge some nutrients to supply the bacteria. The bacterium thus "blackmails" its host into supporting it.

What does the individual bacterium gain from emitting the toxin? The starved bacterium can probably feed on nearby body cells the toxin disintegrates; perhaps the toxin also protects the starved bacterium against other diphtheria bacilli nearby, who are just as starved as it is (see the discusssion on antibiotics in chapter 14). If the host does not cooperate, it is to the parasite's advantage to act virulently; it is only the host's support that enables the nonvirulent phenotype to overcome the virulent one.

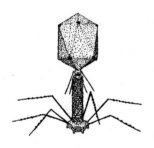

In diphtheria, this triad of host, nonvirulent parasite, and virulent parasite involves different phenotypes rather than different genotypes of the parasite. It may well be that similar relationships occur between hosts and various strains, or genotypes, of fungi, bacteria, or viruses, some more virulent and some less so. Here again, the host can assist the nonvirulent parasite and use it as a partner against its virulent relative. Reliable communication is as necessary between such partners as it is between any others.

THE IMPLICATIONS OF ASSUMING A STATE OF EQUILIBRIUM

To sum up, our approach to parasitism is different from that of researchers who assume that they are witnessing "arms races" between parasites and hosts. As we have stated, any adaptation—including any social system, whether it operates within the species or between species, as parasitism does—can be explained away as not "yet" having reached its logical evolutionary endpoint. Such assumptions, by their very nature, can accommodate any findings and do not generate new predictions or open new avenues of research.

Our own research hypothesis regarding parasitism is always based on the assumption that what we see today is a dynamic state of equilibrium, a balance that results because each party is doing its best to reproduce as successfully as it can. In cases where we do not understand this equilibrium, we assume that we have not yet found all the elements of it, and that we have to look for the missing ones.

Just as the assumption that animal communication is necessarily reliable opens one's eyes to the richness of communication actually taking place in nature, so also the assumption that social systems are in a state of equilibrium enables one to see and appreciate the incredible complexity and beauty of these systems. This approach leads us to ask new questions and devise new hypotheses to test, and so ultimately it turns out to be far more fruitful than its alternative.

INFORMATION CENTERS

FOOD SOURCES AND SOCIAL ORGANIZATION:
THE WHITE WAGTAIL

The white wagtail is a common winter migrant in Israel. Its social behavior varies.[1] Some individuals wander around in flocks and look for such sporadic sources of food as a tractor plowing or reaping or a herd of sheep grazing and stirring up insects. Others find a more permanent source of food, such as a barnyard, a collection of trash containers, the corner of a landfill, or a vegetable garden and establish a territory that they defend against other wagtails. Each territory is defended by one male—but not necessarily the first to stake a claim, since stronger males often expel weaker ones and take over their territories.

In some of the territories, a female joins the male. The two behave as if they are going to establish a nest together, but they do not copulate. The female helps defend the territory and chases away other females. Sometimes the relationship between the two lasts all winter, but often it does not. Sometimes the female moves on to another territory or rejoins a flock; sometimes she is expelled by a stronger female; and sometimes one female establishes a relationship with more than one male and feeds in both their territories. Males, too, leave their territories if their food sources dry up; they then rejoin flocks.

In fact, we have found that individuals will claim territories for themselves or form flocks depending on the way we distribute food in the field. We tagged members of one flock with colored leg rings. When we then presented that flock with small heaps of food, each heap was taken over by one individual, who chased others away from it. But when we scattered one such heap over a larger area, an individual could no longer defend it effectively; he stopped trying to and fed side by side with others. When we stopped supplying food, *all* the birds who held territories went back to the flock. As far as we know, it was the first time that an experiment in the field showed that environmental factors can change social behavior.

As much as the social behavior of individual wagtails varies, these birds have one thing in common: every evening, all wagtails gather in communal roosts— sometimes thousands of them together. Both flocks and territory holders congregate in the roosts, and some flocks fly fifteen to twenty miles to join them.

COMMUNAL ROOSTS AS INFORMATION CENTERS

Communal roosts are not unique to wagtails. In winter, starlings gather by the millions to roost in woods and reed thickets; some of them travel seventy miles every evening to join the roost. Sparrows, egrets, gulls, swallows, crows, doves, and many other species gather to roost together. What do birds gain from making the long daily commute to the central roost?

One might think at first that the birds select a resting place that is protected from predators, say, in the heart of a city or a swamp; or perhaps one that is warmer than its surroundings, such as an urban area or a warm valley; or perhaps a spot sheltered from the wind. But warm, protected places are not scarce, and one need not travel great distances to find them. Some have suggested that the communal roost itself is a defense against predators; but for that purpose a few hundred birds would be enough. Why, then, do birds gather in such huge numbers?

Communal roosts can be seen during the day as well. Ward,[2] who studied the huge midday roosts of black-faced diochs, suggested that they enabled individuals who had not found enough to eat that morning to join a flock that had located food. Such a theory was first proposed by Darling[3] to explain the value of breeding colonies: he suggested that gulls take their cues from those neighbors they see returning to the colony with food; when the successful providers go out again to forage, the observers can then follow them to their sources.

When we first tackled the issue with Ward,[4] we began by considering the species who congregate in large roosts. We found that those who do tend to search for food that appears in short-term concentrations, such as schools of fish or swarms of insects, or in unexpected places like freshly plowed fields or fish ponds that are being drained. For individuals of such species, it is sensible to make use of the knowledge of others. The food of species who tend to roost alone or in small groups, on the other hand, is generally spread sparsely over large areas in quantities that do not suffice for more than a few. For such species, there is no point in following others to a food source, and thus no point in communal roosts.[5]

Since we first published the idea that communal roosts serve as "information centers" in 1973, many studies and experiments that support it have been done.[6] The theory is now widely accepted. Brown observed swallows coming to their nests in nesting colonies with food in their beaks, and other swallows following them when they went back to get more food.[7] Heinrich studied crows in the northeastern United States.[8] In winter, these crows feed on carcasses they find on top of the snow. When Heinrich put such carcasses in par-ticular places, he found that the morning after a single crow discovered them, many crows came to feed on them. He noted that these birds flew straight to a carcass they had not known of previously; most of them came directly from their communal roost, following the roostmate who had found it the day before.

In our original article on roosts and gatherings, we suggested that all gatherings of animals evolved to provide individuals with information about food sources. Today it is clear that some gatherings provide other benefits.[9] Animals also gather to find mates: the communal leks of ruffs and of black grouse that were described in chapter 3 hold hundreds, sometimes even thousands, of birds, and the large concentration of males enables females to choose the best of them. Wagner found that razorbills—a monogamous bird species in which both mates take care of offspring—gather in special mating arenas to copulate both with their mates and with others (extra-pair copulation). It would make sense that gathering together in breeding colonies and in communal roosts, whether during the day or for the night, also gives birds the opportunity to get to know potential mates and thus to choose their mates better, though there is no proof of that yet. Some breeding colonies primarily offer protection from predators, and yet other gatherings, usually small ones, enable individuals to pool their body warmth on cold nights.[10]

There are many additional examples of information centers. Sandgrouse, for instance, are birds of the steppe and desert; they are about the size of pigeons, and they eat grains and greenery, which are distributed very unevenly in their desert habitat. In the summer they fly large distances to sources of water. Some species drink in the morning, about an hour after sunrise, and others drink in the evening. The sandgrouse come to the water sources in couples or small flocks; once there, however, they gather into large flocks and spend time together chattering, chasing each other, courting, and preening their feathers. They are very cryptic in color, and the flock looks like a field of stones—which suddenly turn into birds when they rise and fly together to the water. When the sandgrouse rise up in flight they are visible and audible at a great distance. No doubt these gatherings, which can be seen in several places in Israel's Negev desert, serve as information centers.[11]

INSURANCE AGAINST EVIL DAYS: WINTER GATHERINGS OF ROOKS

Studies of rooks at Aberdeen, Scotland, seemed at first to contradict the theory that communal night roosts are information centers. Researchers at Aberdeen University have been observing local rooks for decades. Rooks are crowlike birds that nest in communal rookeries, from which they disperse to seek food. In winter, rooks from several rookeries congregate together and form communal roosts in marshes. In the morning, each flock flies to the trees in which it nests in the spring, stays there for one or two hours, then roams around seeking food. In the evening the rooks return over long distances to the communal night roost. Even though they spend the night together, members of different rookeries were never seen feeding in each other's areas. Moreover, the food rooks eat is more plentiful in winter than in summer, but in the summer the different rookeries do not gather in the communal night roost.

I suggested to Feare, who was then a student at the research station at Aberdeen, that when unusually severe weather strikes, the rooks might lose their regular

food sources; if that should happen, information provided at the central roost by members of other rookeries might prove crucial. As it turned out, the next winter was exceptionally harsh. In the past, researchers had not gone out to watch rooks in foggy weather. This time, Feare did—and he was rewarded with a sight never before seen by members of the Aberdeen research station: a flock of rooks from one rookery joining a flock from another rookery in the daytime. They all flew together to a farm where a pile of corn jutted out of the snow, and there they fed.[12] The snow and fog continued for several days. The usual food sources exploited by the rooks were covered with snow, and the fog made it hard for them to find new ones. They might have starved to death had they not followed the members of other rookeries, who knew where to find food. Because they made the daily investment in the long-distance flight to the communal night roost, the rooks had insurance against emergencies.

FLOCKS AND LONERS: THE COMMUNAL ROOST OF KITES IN COTO DOÑANA

An individual can get other important information about food by joining a night roost. Heredia, Alonso, and Hiraldo[13] studied red kites in Spain. In their research area, Coto Doñana, the kites feed on chicken carcasses. A carcass, once kites discover it, doesn't last long, and they are widely scattered; by the time kites gather in the night roost, carcasses found that day have already been eaten. Even so, most kites leave their communal night roost together and fly over the marshes and sand dunes in a scattered flock. Once a kite finds a carcass, the others join it and fight over it.

Interestingly enough, the flock the researchers observed was composed mostly of migratory, wintering kites. Local kites, who knew the area well, did not as a rule come to the roost and did not join the flock but rather went out on their own, usually in their regular territories. The researchers also found that kites that had fed well the previous day also preferred to strike out on their own, apart from the flock.

A group of kites can find an isolated chicken carcass more easily than a single individual can. But once a carcass is discovered, its finder has to fight for it. Very hungry kites, for whom even a mouthful would make a difference, do not have a choice: they go with the flock, which is likelier to bring them to a food source sooner, even at the price of having to struggle at that point with their flockmates to eat some of it.

On the other hand, a kite who knows the area well or who is well fed can afford to search longer and then eat at its leisure, far from other kites. Only if it does not find food over a stretch of days would it benefit from joining the flock. The communal roost, then, enables kites to search for food in a group when they need to; in addition, by checking the number of other kites who join the flocks, an individual can estimate how abundant food is in the general area. It can then decide whether to stay or move on to another region.

BRIGHT ADULTS AND DULL YOUNGSTERS: HANDICAPS IN FOOD SQUABBLES

Gulls also search for food in a scattered flock, and like the red kites, they often fight over what they find. A bright white gull diving at a piece of food is visible from a distance and captures the attention of other gulls, who know that a diving gull means a meal. But the summons may well end in a squabble, and there is no certainty that the one who discovered the food will get to eat it. A gull must both find food and defend it.

The gull's bright color helps other gulls but is clearly a burden to its bearer: if it were not so visible from so far away when it dived, the gull might enjoy its prize in peace. And indeed, only adult gulls are so conspicuous. Young ones' backs are a dull gray-brown, probably for camouflage, and their underparts are grayish rather than brilliant white. If a particular gull grew to adulthood without losing the dull, cryptic colors of its youth, then while it could track its bright white fellows to see where they found food, it would not disclose its own resources. One could say, then, that adult gulls are altruists, and the young are selfish. But we say, of course, that the individual who bears the brilliant white feathers must benefit by doing so or the trait could not have evolved.

What is the point of the adult gull's bright plumage? Again, the Handicap  Principle provides an answer. Gulls nest in colonies; this means that they have to be able to find food in the presence of other gulls and to defend it successfully. A gull looking for a mate has to make sure that the mate it chooses is able to stand up for itself. One who is bright in color shows off the fact that it has managed to compete successfully with others who can see its every move clearly; one who has a youngster's cryptic coloring does not provide that proof, and indeed until it loses its dull plumage, a gull is far less likely to find a mate and raise a family. The same principle applies to certain other bird species that feed in loose flocks. Egyptian vultures, for example, gather to feed on carcasses and on piles of garbage; the adults have striking colors—black and white in this case—while young individuals are a dull, cryptic brown.

HOW INFORMATION REPOSITORIES WORK

Wagtails in a night roost that draws birds from a distance of some fifteen miles pool together information about the availability of food in an area exceeding 400 square miles. In a night roost of starlings, which draws birds from up to seventy miles away, information about food in an expanse of more than 12,000 square miles—a huge store of information, by any measure—is available.

It is quite likely that the birds who come to such a roost learn more than simply where to find food the next morning. For example, migratory birds must decide when to move on. The number of birds who need help locating food in the morning is an indication of how available food is in the area. A starling that arrives at the night roost hungry probably waits a while the next morning and watches others going off to their sources of food. When resources get thin, one would expect the number of morning loiterers to rise, and that very increase can tell other starlings that it is time to move on to a new region.

Once a starling notices that the total number of birds coming to a night roost is down, it would make sense for it to join one of the flocks migrating to other regions. Even if the individual bird still knows of good food sources, it will suffer if they dry up later; finding new ones will be difficult on its own or in a small flock. In unfamiliar places, large flocks that cover wide areas can find food more easily—especially if some members of the flock know the area from years past.

Bucher recently suggested that the famous passenger pigeon (*Ectopistes migratorius*) became extinct partly because its traditional roosts were disrupted by hunting and development, and the reduced populations could not gather the information they needed to find the mast crops of breech trees, the scattered and unpredictable food supply on which they depended.[14]

Wynne-Edwards described the gathering of birds in communal night roosts and suggested that such roosts evolved in order to enable a population of birds to spread itself over an area according to the distribution of food in it. We do not accept Wynne-Edwards's explanation of the evolution of communal night roosts, but it should be noted that the collective information in these centers does indeed indirectly help adjust the distribution of birds according to food distribution.[15]

HUMAN GATHERINGS

Human gatherings, too, serve as information centers. Immanuel Marx followed festive gatherings (zaharas) at saints' tombs among Bedouins in the Sinai Desert.[16] He found that when the political situation was stable, few came to such gatherings; when conditions were volatile and the future uncertain, the number of participants rose markedly. Whether or not it is the main reason the festivities are held, it is clear that one function of the gathering is to provide participants with information.

Amotz himself found out how beneficial communal prayer can be. He was called up for army reserve duty. None of the soldiers in his unit knew what they had been summoned for, where they were going, what they would do there, or when they would be released. The unit was taken in the dark of night to an unknown location, and the soldiers went to sleep. In the morning, the pious among them rose early and went to prayer, while the others stayed in bed for another precious half-hour.

Amotz woke up among the latter, who were all as clueless as they had been the previous night. But the guys coming back from morning prayer knew everything. The congregation had included men from other units as well—military intelligence, transportation, operations, and so on—and chats before and after prayers left the devout fully informed about the unit's prospects. The need for knowledge is one of the major reasons people gather in clubs, pubs, and other places[17]—even though what gets a person out of his or her house may be the wish to go to a sporting event, watch birds, have a drink, or fulfill religious obligations.

Veblen, who lived at the end of the nineteenth century, explained why printing press workers were particularly keen on drinking contests at that time: they had a specialized skill that with the rapid expansion of publishing was in great demand, and thus they could move from one place to another and work for whoever would pay the most. These wandering workers made transient connections with people they had not known before, who were their professional peers and occasional competitors. By drinking together, they could get to know each other and brag about their abilities to their new friends. Thus, bars enabled each to establish his social and professional status and probably also served as information centers for those seeking work.[18]

COMMUNAL DISPLAYS AT GATHERINGS: PROMOTING THE ROOST OR MUTUAL TESTING?

One of the typical features of a communal night roost is the attention-grabbing behavior of its participants: sparrows and starlings send up a loud chorus from the tree they have chosen to roost in; egrets and wagtails stand on high, prominent perches. Breathtaking aerial maneuvers are also characteristic of such congregations: flocks of starlings, wagtails, or swallows who have gathered together rise suddenly and tightly circle the roost, making sudden, sharp turns and climbing and plunging abruptly before descending again into the tree.

Wynne-Edwards suggested that the purpose of such showing off is to advertise the site of the roost, to the benefit of all: the more birds at a site, the more information available there.[19] But this, again, is a group-selection argument. It may be best for the group if the site is advertised, but that does not explain why each individual should invest in the advertising, since those who do not take the trouble still benefit from the body of knowledge assembled through the efforts of those who do.[20] Another theory is that flocks perform flight displays to confuse raptors, and indeed flocks approaching communal roosts tend to react to raptors by executing sharp turns. But the aerial maneuvers above communal roosts are too many, too lengthy, and too elaborate for this to be a convincing explanation.

Perhaps each bird participates in such maneuvers to compare its own ability with that of the flock.[21] It is reasonable to assume that some individuals are more skilled in flight than others. Ideally, the individuals in a flock will be of comparable ability, so that, when necessary, they can fly in harmony. If one is significantly less skilled than the rest of the flock, then when there is danger, one may find oneself alone, exposed to raptors,[22] or may lag behind in the search for food. On the other hand, if an individual's ability is much greater than that of others in the flock, it may miss out on rich sources of food that weaker birds might not reach. Perhaps by testing itself against others in aerial maneuvers, each individual decides which flock it should join the next day.

It is difficult for researchers to follow individual behavior in a huge congregation of birds. Only when technology enables us to mark many individuals and track them closely within such a large gathering will we begin to understand the mechanics of flocks better. Still, we believe that individuals in flocks and in roosts are aware of each other's doings, and that many of them know and remember one another. Although we don't know exactly how individuals benefit from calling, performing aerial maneuvers, and so on, we do think it possible that showing off "in public" increases an individual's prestige in flocks and communal roosts as it does in other interactions among animals. As always, we are guided by the logic of individual selection—even when seeking to understand remarkable phenomena of collective social activity like roosts and flocks.

PART IV

HUMANS

HUMANS

H uman social life, like that of all other organisms, reflects the interaction of cooperation and competition among collaborators. We don't mean to suggest that human social systems are not vastly more complex than those of animals. Still, we believe that the same principles guide both; the behavioral mechanisms that survive over generations are those that increase the number of the individual's childbearing offspring. Thus, we will be guided by the Handicap Principle as we examine the logic behind mechanisms of social behavior and methods of communication among humans.

Some people object when human behavior is compared to that of animals. Yet we do this routinely in human physiology. Research on the heart and circulatory systems, the kidneys, and the immunological systems of animals has taught science a great deal about the functioning of the human body. Why, then, should we not seek the same kinds of insights when we consider body parts and traits that serve social purposes?

• • •

INNATE BEHAVIOR IN HUMANS

One example is the taboo against incest. In most human societies, marriage between close kin—between parent and child, or brother and sister—is avoided or forbidden. These rules are usually considered a part of the local moral or religious code. But whatever the ostensible reason for them, these taboos prevent inbreeding that might cause genetic damage to offspring. Animals too commonly avoid incest.

It is interesting to see the mechanisms that serve to prevent such behavior, both among humans and among animals. Babblers, for example, avoid mating with individuals that they helped rear or ones that were present in the group when they were fledglings. This is not a perfect mechanism, since it can prevent a babbler from mating with one who is not really its relative, but who happened to be its groupmate when it grew up. It turns out that unrelated persons who grew up together from infancy in the same communal children's house on an Israeli kibbutz seem to avoid marrying one another, although they are perfectly free to do so.[1] In more traditional settings, growing up so close together almost always meant growing up in the same extended genetic family. It seems that among humans, too, there is an unconscious mechanism that prevents close relatives from mating; legal and religious prohibitions only give expression to and reinforce inborn behavior that evolved through natural selection.

Another unconscious mechanism operates when we size up strangers. Hess conducted an experiment with two identical photographic portraits:[2] he retouched the photographs to make the subject's pupils a tiny bit larger in one and a tiny bit smaller in the other. He then asked people who did not know what he had done to describe the person in each portrait. When shown the portrait with the dilated pupils, they consistently described the subject as a nice, pleasant person; when shown the other portrait, they said the subject looked vicious, or dangerous. This clearly demonstrates our faculty to collect and evaluate information without being aware either of the process or of the reasons for our responses.

Why should pupil size make a difference? A contracted pupil, like a contracted aperture in a camera, increases the sharpness of the image. A person who intends to attack needs to see the opponent very clearly and to keep the opponent's image in sharp focus as they move toward or away from each other—this is what photographers call a "high depth of field." A person with no intention of attacking can afford to see less clearly and to let another's image become "softly" focused; to such a person, the handicap involved is trivial, while an attacker would find it unaffordable. One can assume that a person with dilated pupils is not going to attack. The same study showed that when people are at ease with individuals they love, their pupils dilate. Thus, unknowingly, we provide onlookers with reliable information about our emotions and intentions.

We are all familiar with the difficulty of reading or driving with dilated pupils following an eye checkup. The ophthalmologist uses belladonna drops to dilate the pupils. In Italian *bella donna* means "beautiful lady"; women used to apply belladonna drops to their eyes before going to a ball, in order to make a doe-eyed impression and arouse the affections of suitors. They didn't consciously realize what it was that made that impression "count" in the eyes of their beholders. The handicap they took on—their dilated pupils—made it difficult to see their suitors very clearly, or evaluate their intentions well.

THE HUMAN BODY AND ITS DECORATIONS

Hair

Like most animals, humans have body parts that serve as signals. The most prominent of them is the hair on the scalp. Humans are the only animals whose hair grows beyond a certain finite length; though there are some human populations with short hair, the hair of some people may even reach the length of the body itself. What message can be expressed by long hair? What handicap does it entail, and why is it unique to humans?

Two of the important traits that distinguish human from ape are a higher level of intelligence and hands capable of fine manipulation. Long hair emphasizes both these traits. Without dexterous hands to tie, braid, or cut one's hair, and without the understanding that long, unkempt hair impairs one's vision and movement and should be taken care of, one could not survive with constantly growing hair. No other animal has the mental and manual dexterity to make long hair into an asset rather than a liability.

Ancient humans probably lived in family groups, each with a territory that it defended against neighbors. As a result, humans often had to assess potential rivals from afar. Even more important, since marriage between close kin could be detrimental to the offspring, people had to find mates in other, rival family groups— and to evaluate potential mates from a distance, without revealing themselves prematurely. Hair can provide considerable information. A well-kept and well-dressed head of hair testifies to the amount of time its owner can afford to spend taking care of it; it can proclaim, both at close range and at a distance, the patience, skill, and imagination of its bearer. The hair's condition and luster also bear witness to its owner's physiological state and general health.

Eyes, Eyebrows, and Eyelashes

Eyes, eyebrows, and eyelashes convey a wealth of reliable information in the course of social relations. We are so used to watching the eyes of people we interact with that it makes us feel uneasy, even threatened, to talk with someone wearing sunglasses; and indeed, wearing sunglasses indoors, where the eyes do not require protection from the sun, is considered unpleasant and impolite. The size of the pupil, as we have seen, conveys information about friendly or hostile intent. The colored iris and white eyeball make the movement of the eye easy to follow and reveal the direction of a person's gaze, and thus the object of interest; the eyebrows accentuate the direction of gaze, making it easy to discern from farther away.

Eibel-Eibelsfeldt[3] observed that, in most human societies, raised eyebrows convey friendly greeting, and puckered eyebrows a threat. Puckered eyebrows, like contracted irises, indicate the sharply focused stare of somebody gathering information—quite possibly with ill intent. Raising the eyebrows involves some loss of focus—try to read these lines with raised eyebrows and see. This cost is small to an individual who is honestly friendly but heavy to a cheater who wishes to signal friendliness while intending to do physical harm; the signal is therefore difficult to fake. Again, the signal, though reliable, is often unconscious.

Men's eyebrows are usually thicker than women's, and in both genders they get heavier with age. What is the purpose of thicker, heavier eyebrows? In human society, as in that of many animals, the older adults are in charge. A dominant person—just like a dominant animal—benefits from displaying intentions clearly; this way, others can accede to the dominant's wishes and avoid conflict. A subordinate, on the other hand, has something to lose by disclosing intent too clearly, because a dominant may intervene and prevent him or her from carrying out that intent. Subordinates, then, benefit from probing their way carefully and finding the path of least resistance. This is probably the reason why younger persons tend to have fine, inconspicuous eyebrows: such eyebrows still help display their bearers' intent to nearby observers, but they draw less attention to it.

Lowered eyelids and blinking also serve as signals. Lowering one's eyelids means giving up the collection of visual information—which an individual who is poised to attack or to mount a vigorous defense cannot afford to do. A brief glance followed by lowered eyelids therefore signals nonthreatening interest, a well-known come-hither look. Painting the eyelids increases the distance from which others can see their movement. Blue is best for this purpose: it is different enough not to blend with the eyeballs or the rest of the face, and reflective enough to convey the finer details of movement.[4]

• • •

Nose and Facial Wrinkles

Noses and facial wrinkles also convey emotions and intentions, whether friendly or not. The longer the nose and the deeper the wrinkles, the easier it is for an observer to deduce intentions. A wrinkled face with prominent features is considered "expressive" and "communicative," more so than the smooth face and short and rounded nose of a baby or a child. An adult with a smooth face that expresses few emotions is said to be "baby-faced"—not a wholly complimentary term. The famous actress Sophia Loren rejected in an interview the idea of cosmetically stretching her facial skin to eliminate wrinkles; she would sacrifice her expressiveness, she said.

A nose makes it easier to tell what direction the person is looking in. Indeed, busybodies are said to "stick their noses" into others' affairs. Like eyebrows, the human nose gets larger with age, probably for the same reason—to enable more dominant individuals to display their intent more clearly. Makers of masks are well aware of the effect of noses: witches and evil persons are given long noses, while clowns, who are harmless, are given rounded ones. The rounded nose does not point directly at any particular person and enables the clown to joke and ridicule the audience without insulting anyone in particular. We do not believe that the makers of masks in various cultures are aware of the reasons behind this effect; they simply know, from tradition and experience, how best to convey a particular character.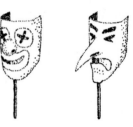

Chin and Beard

Another human body part that is clearly a signal is the beard. Someone showing defiance often raises his or her chin, exposing it to blows, hence the expression "leading with one's chin"—that is, making oneself defiantly vulnerable. A bearded man who does this is thrusting his beard even closer to his opponent. As we saw in chapter 2, a beard can make a man vulnerable in a fight; the Bible itself tell us this, in the story of Joab, who grabbed Amasa by his beard and killed him with his sword.[5] Growing a beard, then, is a way of showing off male confidence. This also explains why women do not grow beards: neither now nor earlier in evolutionary history have women as a rule bragged about their ability as fighters. Handicapping themselves with beards would not serve to advertise any important female trait.

The beard conveys other messages as well. It combines with the sideburns to frame the face like a mane, reducing the face's apparent size.[6] A bushy beard also reduces lateral vision. In older men, graying beards often develop patterns of color that, like wrinkles, increase the expressiveness of the face. The way a beard is groomed provides information about its owner's skills and preferences. And like scalp hair, the beard's color and natural luster provide information on age and on physiological condition.

Red Cheeks and Lips

The red color of lips and the pink of rosy cheeks come from peripheral blood circulation close to the skin. Especially in cold environments, blood vessels in exposed skin cause heat loss, which wastes energy. Indeed, when people are sick or suffering from exposure to cold, these blood vessels contract, and lips and cheeks lose their red color; in extreme cases they turn blue. In cold climates, such as in Scandinavia and Russia, therefore, red lips and cheeks are a reliable signal of health and are considered beautiful: persons who can afford such "wasteful" display prove that they are healthy and vigorous.

Morris suggested that red lips remind men of women's nether labia, which in contemporary human societies are concealed by clothing. But this does not explain why the lips of young children are red, or why cheeks are red.[7]

What about lipstick? Are women who wear lipstick lying about their health? We don't think so. People do not try to hide the use of lipstick, and it is easy to distinguish it from the natural color of lips. The lipstick enhances another function of the color of lips: it brings out the shape of the lips and the fine details of facial expression and makes both clearly visible from farther away. This is why stage actors, who have to project their emotions to the audience, use heavy makeup and strong lines of color to heighten their facial expressions. This is not a message in itself, but a mechanism that enables one to transmit facial messages more clearly and over greater distances.

Menstruation

Body processes, too, can be signals; a case in point is menstruation. Women are unique among mammals in the amount of blood and body tissue they discharge every month in their menstrual flow. This is not required for fertility, for most mammals conceive without it. What, then, is its purpose?

Menstruation is a reliable indicator of a woman's physical condition. When a woman is sick, either bodily or mentally, or when she expends great physical effort, as in sports competitions, her menstruation may cease or become disordered. Pregnancy stops menstruation. Menstruation thus informs a woman's mate or potential mate that she is in good enough shape to bear children—as witnessed by the amount of blood and tissues she can afford to "waste" monthly on menstruation—and also that she is not pregnant at that time, which is very important to one who wants to ensure that his chosen mate's child will be his own.

Breasts and Body Fat

Women's breasts are primarily signals. Most of the breast is fatty tissue that has nothing to do with feeding babies; the babies of most other mammals suckle successfully from nipples on almost flat mammary glands. Large breasts hamper a woman's freedom of movement and waste energy by increasing heat loss. What, then, are they good for?

Large, heavy breasts show clearly that their bearer did not lack for food when they developed: fat concentrated in breasts is easier to gauge than fat spread all over the body. Of course, such showing off is meaningful only in societies that experience food shortages:[8] thus in western society, where food is not scarce, a big, plump woman with heavy buttocks and large breasts is no longer the ideal of female beauty. Breasts also undergo periodic changes in size and texture and provide information about hormonal states and about past childbearing.

Body fat, both in males and in females, signifies affluence even today in many societies around the world. Fat shows that its bearer is a successful person, who has had food in plenty without having to work hard for it physically. In developing countries, this can be important information. In societies where food is comparatively cheap and plentiful, though, like much of modern-day Western Europe or the United States, body fat has lost its value as a signal; it is now considered a drawback. Its opposite, a trim body, shows off in these societies that its owner has enough self-control to eat a sensible diet and to exercise.

Clothing

We saw in chapter 4 how animals' markings accentuate body parts or traits that are important to other members of their species. The same can be said of much human decoration. Clothing, especially when not needed for protection against cold, wind, or sun, is used as decoration and for showing off. Like the markings of animals, styles of clothing evolve when members of a set compete to show off their quality in areas important to that group. The styles that evolve emphasize the points of beauty, proper conduct, and economics that are important to the set and enable observers to compare one member of the set with another. People who do not bother to dress according to the code of their set are also showing off: the message may be that they do not belong to the group and do not care to be judged by its standards, or that their strengths can be effectively displayed even if they don't conform to a dress code.

Ever since property became an important criterion of social success, clothes have served to show off wealth—whether they are made of expensive materials or require a lot of skilled work. Lately, modern production techniques have lowered the cost of clothing to such an extent that it is difficult for most to tell the difference between expensive clothing and imitations. Designer labels are no help, since they are absurdly easy to imitate. It seems that now, more than at any other time in hundreds of years, clothes are becoming less important in showing off. Instead, people show off their bodies, with short clothing, cutoffs, and revealing necklines.

Sometimes standards in clothing are imposed from above, as is the case with army or school uniforms. Of course, such prescribed clothing does not reflect the personal taste or wealth of its wearers, but it enables superiors and commanders to assess differences in personality traits that are important to them, such as neatness and precision. These traits are reflected by the way a uniform is maintained and worn, and they are easier to evaluate precisely because the individuals being compared all wear the same uniform. For the same reason, the first few months at the Israeli air force's school of aviation are devoted to drills, polishing equipment, and precise maintenance of one's uniform. These activities, which may seem to have nothing to do with flight, in fact test each would-be pilot with regard to personality traits that are critical to his or her commanders—orderliness, obedience, and precise execution of commands.

Clothing and other decoration modifies or enhances signals that are conveyed by specific parts of the human body. The neck, for example, supports the head and has to bear its weight. The longer the neck—provided it is also strong enough to support the head properly—the more one is able to move the head around and the better use one can make of sensory apparatus in the head. Men have needed a strong neck for fighting, however, and therefore could not afford a long, thin neck. But women are not usually fighters and can afford to possess and show off a longer neck.

Women, especially young ones, often decorate their necks with close-fitting necklaces or ribbons. A line across a long structure makes it appear shorter than it really is, so it is precisely a *long* neck that one can advertise reliably by tying a ribbon around it; the person who wears the ribbon is proclaiming that, despite the handicap she imposes on herself, her neck can be seen to be longer than those of her competitors. A girl with a short neck cannot afford to wear a ribbon; such decoration would make her appear ridiculously short-necked.

A slender waist can show that a young woman has probably not yet gone through pregnancy; slim waists and trim ankles are considered beautiful and attractive in women. Indeed, it is difficult for such slender structures to bear the weight of the whole body—and that may be precisely why their slimness shows to advantage the high quality of the woman's body tissues and structures. Both waist and ankles thicken during pregnancy when a woman carries a child, to enable her to support and balance the added weight.

For men, broad shoulders are an advantage: the broader the shoulders, the greater the force available to the arms, other things being equal. Long neckties handicap the display of the dimensions of men's shoulders: the line formed by the tie, which is perpendicular to the line of the shoulders, decreases the apparent breadth of the shoulders—just like the black band on a great tit's breast, as we saw in chapter 4. A man with broad shoulders will appear impressive despite wearing a tie, and even one with unimpressive shoulders can use a tie to show off how much better he looks than one who is even more frail. When attempting to downplay the exhibition of manly strength, men wear a bow tie instead. Good examples are waiters, classical musicians, and European males at a very formal social occasion, where the competition to display masculinity is intentionally reduced.

Some people suppose that one of the functions of clothing is to cover defects, to "cheat" the observer, as it were. However, most articles of clothing that might seem at first to do this turn out to do nothing of the sort. A short woman may make herself look taller by wearing high heels, but her taller rival will appear taller still when she wears them. Likewise, when all wear corsets and lace them as tight as possible, the differences between the thinner- and the thicker-waisted are even more pronounced than before. A style of clothing that is actually designed to cover defects will not be interesting and informative, and it will not prevail. A fashion in dress is accepted, it seems, only if it accentuates an important characteristic in a reliable way.

TESTING THE HUMAN SOCIAL BOND

A person's social life involves a complex set of partnerships and cooperations. At the foundation is the family partnership, formed for the raising of children. In addition to this, one collaborates with others to procure food—ever since the prehistoric communal hunt—for protection, for economic gain, for recreational and religious purposes, and so on. At each level the individual must choose his or her collaborators and must test their willingness to invest in the collaboration and to share their assets. As we saw in chapter 10, the only way to test a social bond is to impose on the individual tested. Not surprisingly, this turns out to be the key to some of the more peculiar behaviors we humans engage in.

When we happen to meet a good friend whom we have not seen in a while, we may approach from behind, slap the person on the back, and utter "offensive" expressions like, "Where have you been hiding so long, you old rascal?" Why are we slapping and "insulting" a good friend, especially after a long separation? Unconsciously, we are testing to see whether this person is still our friend. Only a friend will accept such behavior, turn to us, and greet us happily—as one learns to one's chagrin after behaving this way toward someone who turns out to be a stranger. Only by imposing on and challenging our collaborators can we acquire reliable information about their readiness to collaborate with us further.

We hug our loved ones tightly, sometimes so much so that they have to hold their breath. Only a person who loves us back will agree—will in fact be glad—to bear such imposition. Indeed, all our love signals are impositions of one sort or another: kisses, hugs, and petting intrude on our personal space and impair our freedom of movement. In spite of this, we like them: these behaviors demonstrate that the tester loves us enough to test us and enough even to invite us to impose a similar test on him or her. Even lovers who just hold each other's hand for hours at a time are each giving up the use of a hand for that time, a pretty heavy imposition.

If signals of love were meant only to identify love, clever humankind—*Homo sapiens*—could easily establish conventional signals to replace the repertoire of impositions we use now. But one of the most important issues for lovers is reliability—how sincere is the other? It seems that natural selection could find no other way to test for this information.

Ethologists say that the function of indications of love is to strengthen the bond between partners. If that were true, then one could strengthen a failing bond by holding hands or by reciprocal petting. But when love is gone, such forced closeness will probably only hasten the breakup. We think rather that love signals serve to test the bond. The investment in love signals is very high, especially when the relationship is new and not yet well established, because that is the time when the information is most needed, when one can still dissolve an ill-considered partnership with minimal harm, and when the wrong decision can be disastrous. The burdensome love signals either strengthen the bond or dissolve it; only real lovers accept the imposition.

The Human Sexual Act as a Test of the Bond

Sexual relations for purposes other than procreation are not unique to humans. Some other animals also copulate without regard to the female's reproductive cycle. Porcupines copulate several times every night, all year long.[9] Many monkeys copulate when females are not in heat. The bonobo chimpanzees of the woods of Western Africa have sexual relations all year long, in all possible combinations—male with female, male with male, and female with female.[10] Hoatzins, cooperating birds of the American tropics, copulate during border clashes between groups.[11] Babblers, too, sometimes have sex outside of breeding season. We have even witnessed several cases of a male and a female from rival groups

meeting and copulating under a bush on the border between their territories, even though there was no chance these Romeos and Juliets would have offspring as a result of such trysts.

We think that in all these cases copulation occurs in order to test the social bond between the parties. The imposition involved in sexual relations is greater than in any other signal of love, and it increases markedly each party's knowledge of the other and of their relationship. No wonder that animals who copulate outside the process of procreation are those that maintain long-term partnerships.

In reproductive terms, a stable pair bond is more important to a woman than to a man: the number of children she can have is more limited, as is the time she has in which to bear them, while a man has more chances to pair off with other women, to the possible detriment of his original mate and her children. This may be why it not uncommonly takes much more stimulation for a woman to become sexually satisfied than it does for a man. Anyone trying to "cure" this tendency is acting against the woman's unconscious means of testing the real commitment of her mate.

What about homosexual relations? Any trait that is more common in a population than a chance mutation would be—and homosexuality is clearly such a feature—must have some benefits, since the trait has obviously survived natural selection. This is particularly so if the trait, like homosexuality, seems on its face to impair reproduction. What benefits can come out of homosexual behavior?

As we have seen already, copulation can serve to test the social bond between male and female, and that bond does not necessarily have to do with procreation. For example, some birds, such as stone chats and wagtails, pair up in their winter habitats—not to breed, but rather to defend what is temporarily their mutual territory.[12] In such cases, one often finds that the same courtship mechanism that helps two individuals form a partnership in the breeding season serves them at other times to form a partnership solely for defense. One who doesn't know that these partners have no intention of reproducing and will split up in a few months might think they were going through regular courtship.[13]

Modern Western society sees homosexuality as an alternative to heterosexual relations. But in some societies, especially ones where men and women are strictly segregated, people engage in two parallel sets of sexual relations: those between men and women, which produce children; and those between same-sex partners, which help sustain and enhance companionships of various sorts. In both cases the sexual act can serve as a mechanism that lets the partners test the partnership and sustain it. Just as sex between men and women encompasses more than procreation and serves also to test the bond between them, sex can provide information needed to maintain bonds between partners of the same gender. Again, this is not unique to humans: Trivers describes pairs of female western gulls behaving as perfectly normal couples and together raising offspring conceived through extrapair copulation.[14]

Of course, the sexual act is not necessarily enjoyable to both participants. Just as consideration and care for one's partner can be demonstrated through sex, so can the lack of it. And that lack of consideration elicits another message in response: one who continues to accept such treatment is telling the other very clearly that, unsatisfactory as the relationship is, he or she does not want to end it. That may be because of other perceived benefits that seem more important than care and consideration, or because of a lack of alternatives.

Forced sex—rape—is often used as a means of crude dominance. This is as true among men as between men and women. Again, mountings as a means of showing dominance are not at all uncommon in the animal kingdom, either between males and females or among males: this has been shown in numerous species.[15] And all human languages, it would seem, have some verb that takes on both the sexual and the dominative meanings of the colloquial American English verb "to screw (someone)."

The sexual act among humans shows reliably the quality of the relationship between those participating in it. Between caring and devoted partners, it expresses that care and devotion. An uncaring partner will find it hard to hide the lack of care—which may lead the other member of the pair to leave. The sex act may be used to prove the other's helplessness and powerlessness, as in rape. We will let novelists and poets detail the peaks and abysses of the human sexual act, and all the levels in between. Our point here is a global one: that it is precisely the closeness and invasiveness of the act that make it a means of conveying and receiving detailed and reliable information about the character of its participants and their relationship with each other.

Self-Endangerment in Humans: Suicide as a Cry for Help

We have also seen that animals—nestling birds, for example—endanger themselves to extort care.[16] Among humans, too, weaker partners can sometimes extract more from a powerful partner by endangering themselves. A toddler can force its parent to pick it up and carry it by running toward a busy street, or by sitting in a dangerous place. Some children beat their heads against the wall until their parents give in to their demands. Young children who climb onto furniture and jump at a parent are forcing their parent to catch them. Similarly, women throwing themselves into their lovers' arms could be badly bruised if the lovers did not receive them. All are endangering themselves—most probably unconsciously— and as a result, they get information they would not have gotten otherwise; they are testing the social bond. The risk itself forces the other party to come to the tester's aid and thus to express commitment.

Suicide is an extreme case. One could even redefine successful suicides as unsuccessful calls for help. More often than not, however, death is not the result, and the suicide attempt causes friends and family to give the desperate risk-taker help that was not forthcoming before. The person who attempts suicide may not be aware of this logic: he or she may honestly prefer death to an unbearable life. Such an attempt at suicide is genuine, and that in itself convinces others of the need to help. Sometimes the person attempting suicide may even be half aware of this logic, as when someone takes an overdose of sleeping pills, then calls a friend and says, "If you don't come and save me, I'm dead." The risk is still genuine, though, and it does force the friend to come and help. The risk of death is what persuades others of how desperate the situation is—and indeed, death may seem better than the prospect of a hopeless and helpless life.

HUMAN LANGUAGE: COMMUNICATION WITHOUT RELIABILITY

Human language is unique in the animal world in that it is a system of communication by symbols. It is possible to train animals to understand and process information passed by symbols. Some animals have been taught scores of words. Primates and dolphins can even learn to communicate with people by using symbols, and birds such as parrots and mynas can be trained to produce exact vocal copies of human word symbols, and even to use them in a way that makes sense.[17] Still, as we saw in chapter 6, there is no animal that uses a symbolic, word-based language in nature.

We believe that animals did not evolve a verbal language because the language of sounds and nonverbal communication serves them better. Most animals live in small groups whose members know each other intimately and spend most of their time together. Such individuals share the same immediate surroundings, witness each other's actions, and can discern each other's intentions.[18]

For example, when a babbler makes a begging sound, its comrades do not need to be told that it is begging, or what it is begging for—they know that from the circumstances. What they can't tell is how badly the beggar wants the thing it is begging for; the intensity of the wish is conveyed reliably by the quality of the beggar's voice. When a babbler threatens, its comrades know without being told who is being threatened, why, and what the threatener expects the other to do; again, this is obvious from the circumstances. What they need to find out is how reliable and how intense the threat is. This is conveyed best and most reliably by nonverbal communication.

The information that nonverbal vocalizations do convey is very exact: they express the *degree* of feelings much more precisely than words can.[19] For example, the words *I am angry* do not convey *how* angry one is; to convey the degree of anger by words alone, one has to use more words: "I am very angry"; "I am somewhat angry." Even then, words can express only a few of the infinite gradations of anger that are possible; but nonverbal vocalizations reflect such gradations admirably.

On the other hand, human beings who are not familiar with one another may perceive a given circumstance differently. A person listening to a stranger may be unable to relate the intensity of vocalization to the degree of emotion, as the stranger's constant companions can, from past experience. This is especially true in meetings between people of different cultures. In that situation, it is best to communicate with words, even if the parties have to use a dictionary to translate from one language to the other. Verbal language may be a poor, inexact medium for expressing feelings and degrees of feeling—but it can prevent the misunderstandings that might arise from incorrect interpretation of nonverbal vocalizations.

We don't know how symbolic word language evolved in humans. But once it did, it enabled groups of humans who were not each other's constant partners to collaborate temporarily: for protection, for war, for hunting. These alliances were short-term and required cooperation among partners who did not work together most of the time. Such partners also had to be able to discuss things that were not within view—to say, for example, "There is a saber-toothed tiger on the other side of this hill," or "You come up the valley, I'll hide in the canyon by the spring."

People who are accustomed to working together act and communicate differently than those who do not as a rule work together. A regular crew of movers do not need many words when it comes to carrying a piano up a flight of stairs. Each of them knows from experience where to grab, how to carry, when to push, pull, lift higher, or stop. All they need is a signal at the instant they are to lift the piece off the floor, and possibly a few monosyllables at certain moments along the way. A few short grunts by the leader may well be sufficient.

On the other hand, if a group of people who have not worked together before are called upon to lift a heavy item, they will have no choice but to discuss each move in detail before attempting it and to continue the discussion while carrying out the task. Even then, their teamwork will be less perfect than that achieved by the first group by means of a few short grunts. But this second group does not have enough experience to work as a team with few or no words. Verbal language is the only tool that can enable them to labor together successfully.

The rub is that verbal language does not contain any component that ensures reliability. It is easy to lie with words. There is no substitute for the reliability and precision of nonverbal vocalizations, as we saw in chapter 6. This is why, even after humans evolved verbal language, no human society ever gave up the use of nonverbal communications.

DECORATION, ESTHETICS, AND THE EVOLUTION OF ART

As we saw in chapter 4, decorative markings spread in a population when individuals who are so decorated are preferred as mates over ones who are not. This preference results when the markings in question help observers reliably identify the better individuals. It follows that those who, in effect, select the pattern, have to have some preexisting yardstick to assess the decorated individuals. This measuring stick must enable them not only to distinguish the good from the bad, but more important, the better from the merely good. In humans, this capacity expresses itself through what is generally called sensitivity to beauty, or esthetics.[20]

Obviously, the markings of other species were not selected by humans; rather females and males of these species chose as mates individuals decorated with the "right" markings, which caused the evolution of this specific decoration. Still, highly decorated animals impress us as being beautiful; this is because the decoration is not random but rather tends to be symmetrical and appropriate to the animal's shape, having evolved to emphasize specific traits by which these species have adapted to their environment and way of life. Animals, too, show a sense of esthetics: for example, they prefer symmetry and completeness to the opposite—whether in a peacock's tail, or a bowerbird's bowers. A nice example is the pair of long, highly decorative feathers of the king-of-Saxony bird of paradise, which are valued as decorations both by bowerbirds and by tribespeople: the selection process by which these feathers evolved was carried out by female birds of paradise, not by humans or bowerbirds—but all three species find it attractive.[21] Birdsong sounds pleasing and harmonious to us; it may be that birds prefer such songs because these enable them to more easily tell the ability of the performer.

Neither the beauty of other species nor human appreciation of that beauty, however, constitutes "art." Art is the purposeful decoration of objects, and indeed the production of objects whose only function is an esthetic one. Art has obviously been practiced by humans from prehistoric times. Yet how did it evolve? It is highly improbable that art sprang forth fully formed from the brain of some mysterious prehistoric human who decided to express an admi-

ration for nature, or who contrived to charm animals before a hunt by rendering their likeness in a cave painting such as the ones that survive to this day.

These cave paintings show a very high level of artistic accomplishment. The craft of painting is not a trivial skill: it demands an ability to see and an ability to execute that are both difficult to achieve. The skill of putting a line or a dot in the right place could evolve only if from their very first attempts, aspiring artists were rewarded for developing that skill. Such reward could only come from others—but why should others pay attention to such lines and dots? Whatever humans gain by appreciating and creating art, its value must have an actual and concrete basis: to claim that its value is "spiritual" or "esthetic" is simply to miss the point, since such "feelings" and "spiritual needs" too evolved for specific reasons. How, then, could people have benefited by their very first attempts at art?

We have seen that wrinkles and eyebrows increase the distance at which facial expressions can be seen. People may have noticed that a smudge of dirt in the right places on a face enhanced that effect. The first step in the evolution of human art may have been taken when people began experimenting with such marks and then deliberately applying them in the right places. The more skillfully such lines and dots were placed, the more valuable, and valued, they could be—rewarding the better of the budding artists.

Once the production of artifacts evolved beyond its very crude beginnings, people must have noticed that certain decorations could bring out the quality of the objects they made. For example, as we saw,[22] a circle in the center can make it easier to distinguish a perfectly round object from one less perfect. Without a compass, it takes a good bit of skill to make, say, a perfectly round plate. Artisans

who had such skill benefited by decorating the center with a circle that helped show off the perfection of their work.

Decorating the object itself demanded skill, however—if the circle was even slightly misplaced it would distort the perceived shape of the object. An outline can also show off clearly the quality of a crafted item, by making it impossible to hide imperfections in its edge—but only if the item is outlined perfectly. A circle that is not exactly in the center or an outline that is crooked makes the product seem *less* good than it is. It may have taken humanity a long time to evolve the talent to use and arrange shapes and lines in effective, satisfying ways, but at every stage of that evolution the more talented artisan could be recognized as such and rewarded.

As they learned to make new artifacts, then, people also learned to decorate them in a style that accentuated their quality. Decoration that does not bring out the perfection of shape, material, or craft is considered "tasteless" or "kitsch." In other words, we notice when the style is not appropriate to the form or the medium. To take a modern example, heavy cutwork that brings out beautifully the quality of material and craft in fine crystal looks tacky when copied in molded plastic. Plastic, though, looks beautiful in designs that bring out its own unique qualities. Not all have the skill to create such designs and execute them well. People who do are artists and talented craftspersons who understand and know how to work with the material they use. This ability could have been—and apparently was—appreciated from the very start—from the first time a smeared-on dot of colored mud was applied effectively.

ALTRUISM AND MORAL BEHAVIOR

To many people, the difference between humans and animals is that animals act through instinct, to advance their material interests, while humans have spiritual and moral drives that are lacking in animals. Yet among babblers and other animals we saw some behaviors that, if we found them in humans, would be considered evidence of high moral standards. The babblers show a lot of consideration for their companions. They share their food with other babblers; they come to the rescue of their fellows, endangering themselves; they spend time acting as sentinels for the group; they feed young that are not their own. They also refrain from incest and don't copulate in the presence of other babblers. These "altruistic" and "moral" acts, as we believe, increase the overall success of the individual who performs them—and that is precisely why babblers do perform them.

We believe that among humans, too, there is a correlation between acting morally and ethically on the one hand and success in life on the other. We believe that, other things being equal, those who behave according to their society's moral principles—those who can afford to do so—are likelier to succeed than those who do not. What, then, is there to distinguish the moral and ethical good from the material good?

We think that the distinction is artificial, based on our limited understanding of behavior. For example, an altruist is defined as one who assists another without expecting any payback. But a gain or a benefit may come in a form other than material payback. Altruistic acts obviously demonstrate—and are perceived as demonstrating—the abilities of those who perform them. Not all of us can afford to give away part of our money or other possessions, or to risk ourselves in order to save another; and among those who do these things, some do them better than others. Investing in another's welfare shows off the altruist's quality, improves his or her social standing, and increases his or her chance at success. True, some altruists lose more than they gain, particularly when they volunteer to risk themselves for their comrades or their country; but more often, the altruists return from the front lines with honor and laurels—having improved their own or their children's chances, or both.

We certainly don't assert that those who volunteer to risk themselves in order to save another are playacting or operating cynically to further their own interests. We have the drive, instilled by natural selection, to occasionally risk our lives for another. But on average, altruists are likely to gain more than they lose. In fact, self-sacrifice is not the only risky behavior typical of humans. Many sports are dangerous—auto racing is a blatant example; so are such pursuits as mountain climbing, voyaging in space, and exploring unknown parts of the planet. But those who succeed gain fame.

Both those who spend resources and risk their health and life for fame, and altruists who invest resources or risk health and life for their comrades, gain prestige according to their society's principles and needs. As a rule, we accord this prestige and respect as a matter of course and do not consider them to be the benefits of altruism. We are much more aware of the cost of altruism than we are of the benefit it brings to the altruist, just as we are much more aware of cases in which a suicide attempt results in death than of the far more frequent cases in which the result is that help that might not have been forthcoming without it is made available.

Patriotism is a form of "altruism" that is hallowed in many cultures. An educational system that stresses love for one's country, and the real need to defend that country, can create an atmosphere in which any attempt to avoid the danger involved brings a loss of social standing. Yet once one reaches the battlefield, the opinion of one's comrades is the immediate motivation. If one asks officers in select combat units what it is that motivates soldiers to risk their lives, they answer that the strongest driving force is shame and the risk that one's comrades will think one a coward. Even mercenaries take risks in battle, although they don't have such a stake in the cause they are fighting for. For them, their comrades' good opinion is all that motivates them to take risks—and this may be enough to move them to deeds of heroism.

Gift-giving is another altruistic act that increases the prestige of the giver. We are ashamed to give gifts of lesser value than those we receive. For this reason, we tend to give our more prosperous acquaintances more expensive presents than we give the less well-to-do, even though the latter might have greater needs. And indeed the custom of shaming rivals by giving them presents exists in many human societies. It was highly developed in the potlatch of the Northwest Indians: they would hold extravagant feasts, at which rivals were presented with expensive gifts; a rival who failed to return in kind lost face.[23] In the highlands of Papua New Guinea, too, some tribal heads lavish gifts on their rivals, attempting to give so much that the other will be unable to reciprocate. Not for nothing did the old Jewish sages say that "the hater of gifts shall live."

Fund-raisers are well aware that donations pledged or handed over in the presence of peers tend to be much larger than donations given in private. This has become a cornerstone of Jewish fund-raising in the United States: charity events are organized so that donations will be publicized as much as possible among the donor's business associates and competitors. Donations are solicited publicly, often at large meetings, by persons of high status, who themselves contribute hefty sums and thus "force" others to do likewise so as not to lose face.

Bakal, who studied charity in the United States, says that these methods were first practiced by Josef Willen.[24] Willen made constructive use of the findings of the American sociologist Veblen, who stressed the importance of conspicuous consumption as a means of showing off economic status.[25] After all, from the giver's point of view, conspicuous donation is simply a form of conspicuous consumption: both prove that one has money to throw away. Showing off one's financial well-being in a reliable manner has a great deal of purely practical value; for example, it can reassure potential business partners and facilitate future deals.

The Jewish sages were well aware that donors gain prestige, just as they understood that recipients of charity lose face; according to Judaism, it is important to give in secret, so as not to shame the recipient. This does not contradict the notion that altruism is a means of gaining acclaim. Donors want to increase their prestige not in the eyes of those who receive their charity but rather in the estimation of their peers, acquaintances, competitors, or mates—who are often aware of the donation even when it is supposedly secret. "Secret" donations are made to protect the self-respect and reputation of the *recipient*—whose esteem is quite likely unimportant to the donor anyway.

When we seem to pursue our own interests, we are considered "selfish" both by ourselves and by others. On the other hand, we believe our tendency to act for others' benefit, which logically does not seem to be self-serving, is an expression of "good moral values"; we feel impelled to follow the path of "altruism," and we respect others for following it. We are vividly conscious of the costs, risks, and dangers entailed—of the handicap involved—and that's precisely why we are impressed by altruism. Yet we consider it bad form to calculate the benefits it may bring us.

Still, the shame we feel when we cannot return favor for favor shows that on some level we are well aware of the prestige altruism brings us, and of the effect this prestige has on our social standing. The wisdom of generations acknowledges that altruism is indeed rewarded—that indeed, as our grandmother used to say, "when you do good, you do well."

EPILOGUE

The Handicap Principle is a very simple idea: waste can make sense, because by wasting one proves conclusively that one has enough assets to waste and more. The investment—the waste itself—is just what makes the advertisement reliable. This idea seemed so obvious to us that we assumed at first that it must already be widely accepted, and so we searched the existing literature for discussions of it.

Our search yielded many previous attempts to explain the waste one sees in sexual showing off. Most of these explanations were very complex, and some were aided by mathematical models, but the Handicap Principle was not among them. And to our great surprise, this idea, which struck us as self-evident, was bitterly resisted by the scientific establishment. Even more to our surprise, this same idea ended up revolutionizing our understanding of communication throughout the living world, up to and including communication within the body.

The Handicap Principle states that the receiver of a signal has a stake in the signal's reliability, or accuracy, and will not pay attention to it unless it *is* reliable. Thus signals are not arbitrary; rather, each signal is the one best suited to reliably convey the specific message it carries. It follows that there must be a logical connection between the message and the signal. The Handicap Principle enables us to make predictions: it makes it possible to figure out from the nature of a signal what message it conveys, and likewise what might logically serve as a signal for a given message.

The Handicap Principle expands our current understanding of evolution in a basic way. Signals, like other traits, evolve through natural selection. But where those other traits are selected because they make an organism more efficient in a straightforward, utilitarian way, *signals* are selected because they *handicap* the organism in a way that guarantees that the signal is reliable. This apparent paradox masks a basic consistency between signal selection and the evolution of other traits: in both cases, traits that spread throughout the population are those that improve an individual's chances of having offspring that in turn will reproduce successfully. This is just as true of a signal that entails a handicap as it is of a bodily structure that makes its owner more efficient.

A theory is valuable only if it can lead to new findings, and in fact, the Handicap Principle has already opened up new avenues of research. A decade ago, many researchers attempted to show how animals use signals to mislead; nowadays an increasing number of studies show that signals reliably reflect the intentions and qualities of the signaler. Even so, researchers do not yet seek the specific cost—the handicap—that makes the signal reliable. Since the handicap—the investment—is what guarantees that a given signal is reliable, it follows that by determining the specific handicap a signal entails, we can better understand the signaler, the environmental conditions, and the message.

Once one accepts the Handicap Principle as a general rule, one can no longer see signals in nature as mere conventions. One therefore has to reevaluate all signals—including those that up to now have been thought simply to identify a given species, age, gender, or any other grouping—all the way down to chemical signals on cell membranes, which traditionally have been seen as merely identifying the cell as belonging to a specific type.

The many examples from humankind that we have used throughout this book show how deeply the Handicap Principle is embedded in human life. The principle can help us better understand many of our natural tendencies, which result from the need constantly to test the bonds linking us with others in various endeavors. The Handicap Principle also shows how the need to cooperate with one's competitors could have led to the evolution of altruism, in humans as in other animals.

Human beings have long marveled at the wonders of nature. The Handicap Principle is one of those theories that show us how nature's intricacies are bound together in an orderly system—a system that is logically simple, that makes sense, and that we can understand. An awareness and understanding of this pervasive order may demystify nature, but it detracts not at all from the wonder we feel when we look about us at the world we live in. Indeed, nature's order may well be its most awe-inspiring marvel.

NOTES

Introduction

1. Zahavi, A., 1975, 1977.
2. Davies and O'Donald, 1976; Maynard Smith, 1976b; Kirkpatrick, 1986.
3. Eshel, 1978a; Pomiankowski, 1987.
4. Grafen, 1990a, 1990b.
5. Lotem, 1993a; Maynard Smith, 1991a; Collins, 1993.

Chapter 1

1. See Maynard Smith, 1965. The difficulty with group selection will be discussed in chapter 2.
2. Zahavi, 1978b.
3. Curio, 1978.
4. Marler, 1955.
5. Sordal, 1990.
6. Morris (1990) describes how man has from ancient times used birds' mobbing instincts to hunt them, using decoys of raptors.

7. Caro (1994) surveys current studies of stotting as well as past explanations of this behavior. See also Hasson, 1991a.

8. Zahavi, 1977a, 1987.

9. Fitzgibbon and Fanshawe, 1988; Caro, 1994.

10. Hasson et al., 1989.

11. Hasson (1991a) recently reviewed communications between prey and predator by pursuit-deterrent signals.

12. Smythe, 1970.

13. Rhisiart, 1989; Cresswell, 1994.

14. Wiklund and Jarvi, 1982.

15. Ritland, 1991a, 1991b.

16. Eshel, 1988.

17. Kruuk, 1972.

18. Eshel, 1978a.

Chapter 2

1. Lorenz, 1966.

2. Maynard Smith and Parker, 1976.

3. Zahavi, 1977a.

4. Ewer, 1968.

5. Baerends and Baerends-van Roon, 1950.

6. Clutton-Brock et al., 1982.

7. Zahavi, 1981b.

8. Morton, 1977.

9. Zahavi, 1982; see more in chapter 6.

10. Davies and Halliday, 1978.

11. Katsir, 1985, 1995.

12. See also chapter 6.

13. Schjelderup-Ebbe, 1992; Lorenz, 1966; Marler and Hamilton, 1966. See chapter 12 for further details.

14. Darling, 1937.

15. Barrete and Vandal, 1990.

16. Lorenz, 1966.

17. A detailed discussion of the drawbacks of the model of group selection can be found in Dawkins (1980). See also Maynard Smith, 1964, 1976a.

18. Lorenz, 1966; Wynne-Edwards, 1986. Williams (1994) uses a different definition of "group selection" and stresses that he does not assume the existence of adaptations that developed "for the good of the group."

19. See Axelrod, 1986; we will discuss this subject further in chapter 12.

Chapter 3

1. Williams, 1966. Williams's idea was further developed by Trivers (1972) who examined its various implications.

2. Jones and Hunter, 1993.

3. Zahavi, A., 1975, 1977a, 1977b.

4. Nisbet, 1973, 1977.

5. Wilhelm et al., 1980, 1982.

6. O'Donald, 1963.

7. See Wynne-Edwards, 1962.

8. Ryan et al., 1982.

9. Clutton-Brock and Albon, 1978.

10. Lambrechts and Dhondt, 1986.

11. See chapter 6.

12. See chapter 4.

13. Thornhill, 1992a; Moore, 1988.

14. Eisner and Meinwald, 1987, 1995.

15. Eisner and Meinwald, 1987.

16. McKaye, 1991.

17. Christy, 1988.

18. Borgia, 1986; Diamond, 1986a, 1986b.

19. Frith and Frith, 1990; Diamond, 1991.

20. Borgia, 1986.

21. Borgia and Collins, 1986.

22. Petrie et al., 1991.

23. Moller, 1994.

24. Smith and Montgomerie, 1991.

25. Smith et al., 1991.

26. Evans and Thomas, 1992.

27. Gibson et al., 1991.

28. Petrie et al., 1991.

29. Hoglund et al., 1993.

30. Gibson and Hoeglund, 1992.

31. Snow, 1976; Foster, 1981.

32. McDonald and Potts, 1994.

33. Van Rhijn, 1973.

34. Hogan-Warburg, 1966.

35. Dominey, 1980.

36. See a general discussion of this subject in Taborsky (1994).

37. Darwin, 1859.

38. Darwin, 1872.

39. See a detailed discussion in Cronin, 1991.

40. Fisher, 1930.

41. See Andersson, 1994.

42. See Mayr (1972) on this subject (p. 97): "Darwin assumed rather naively that 'the best armed males' were also the strongest and that 'the more attractive' males were 'at the same time more vigorous' . . . there is, however, no demonstration of an automatic correlation between the two characteristics."

43. This kind of solution to the problem of waste in courtship computes the evolutionary

value of features by their frequency in a population. Such solutions were later called by Maynard Smith "evolutionary stable strategy" (ESS) and have been used by him to explain many other social phenomena. See Maynard Smith, 1976c.

44. See Fisher, 1930, 2nd edition (New York: Dover Publications, 1958), p. 155: "The possibility should perhaps be borne in mind in such studies that the most finely adorned males gain some reproductive advantage without the intervention of female preference, in a manner analogous to that in which advantage is conferred by special weapons. The establishment of territorial rights involves frequent disputes, but these are by no means all mortal combats; the most numerous, and from our point of view, therefore, the most important cases are those in which there is no fight at all, and in which the intruding male is so strongly impressed or intimidated by the appearance of his antagonist as not to risk the damage of a conflict. As a propagandist the cock behaves as though he knew that it was as advantageous to impress the males as the females of his species, and a sprightly bearing with fine feathers and triumphant song are quite as well adapted for war-propaganda as for courtship." Fisher has no explanation for this effect, and continues (p. 156): "The evolutionary reaction of war paint upon those whom it is intended to impress should be to make them less and less receptive to all impressions save those arising from genuine prowess."

45. See Alcock, 1993.
46. Andersson, 1994.
47. Zahavi, 1981a, 1987, 1991a.

Chapter 4

1. Lorenz, 1966.
2. Wallace, 1889.
3. Mayr, 1942.
4. Smith, 1966, 1967.
5. Katzir, 1981a, 1981b.
6. Snow, 1976.
7. Selander, 1972.
8. Zahavi, 1978, 1981, 1987, 1992.
9. See more in chapter 5.
10. Barlow, 1972.
11. Zahavi, 1978a, 1981a, 1987, 1993.
12. Hailman, 1977; Morris, 1990.
13. Tinbergen, 1953.
14. See more in chapter 8 about the benefits and drawbacks of two colors rather than one.
15. Hamilton and Zuk, 1982.
16. See also Hasson, 1991b.
17. Moller, 1990a, 1992.
18. Watson and Thornhill, 1994.
19. Thornhill, 1992a, 1992b.
20. Parson, 1990.
21. Zahavi, 1993.
22. See Zahavi, 1978a.

23. Petrie et al., 1991.

24. Ridley, 1981.

25. See for example the article by Gibbs and Grant (1981) on changes in beak size that follow climate change and changes in diet in finches on Daphne Island in the Galapagos. See also Weiner, 1994.

26. Lack, 1968.

27. Roper, 1986; Fugle et al., 1984.

28. Jarvi and Bakker, 1984; Norris, 1990.

29. Rohwer and Rohwer, 1987; Rohwer and Ewald, 1981.

30. See Saino et al., 1995.

31. See chapter 7.

32. Hasson, 1991b.

33. Maynard Smith, 1991b.

34. Grafen, 1990a, 1990b.

35. Hendry et al., 1984.

36. This may have happened with ducks: the brilliant plumage of drakes in the Northern Hemisphere shows off their ability to avoid predators despite the handicap of being very visible. Lack (1970) pointed out that on small, remote islands mature male ducks are often drab. It may be that on small islands without predators, where every drake could afford bright plumage, the brilliant colors lost their value as a reliable indicator of the drake's quality, became superfluous, and disappeared. See Zahavi, 1981a.

37. Pond, 1973.

38. Naama Zahavi-Ely, personal communication.

39. See Zahavi, 1987.

40. Borgia, 1966.

41. Hunter and Dwyer, in press.

Chapter 5

1. Redondo and Castro, 1992.

2. Eibel-Eibelsfeldt, 1961.

3. Huxley, 1914.

4. Cullen, 1966.

5. Morris, 1957.

6. Zahavi, 1980, 1987.

7. Simpson, 1968.

8. Boake, 1991.

9. Krebs and Dawkins, 1984.

10. Kruuk, 1972.

11. The assumption that the ability to notice signals precedes the evolution of a signal serves to explain why animals are sometimes attracted to signals that do not exist in their species (Basolo, 1990; Burley, 1986). Ryan (1990) describes many such cases and suggests that when females are preinclined to specific sounds or colors, their sensitivities are utilized by males who adopt these attractive features. He terms this phenomenon "sensory exploitation." But in our opinion, the fact that observers are able to sense one signal or another does not cause the signal to evolve in a particular way. Obviously, no signal will evolve if

the one signaled cannot perceive it. But neither will it evolve if the receiver finds the information uninteresting or unreliable.

Chapter 6

1. Zahavi, 1982.

2. Katsir (1985, 1991) found that the inversion frequency of a babbler call is related to the state of the body making the call: it is very low when the babbler is sitting relaxed on its nest, higher when it is standing in a tree, and higher yet when the babbler is in flight. On inversion frequency in birds and its physiological basis, see Greenewalt (1986) on the physiological side of birdsong.

3. Darwin, 1872.

4. Scherer, 1979, 1985.

5. See, for example, a review by Murray and Arnott on human vocal emotion (1993), which was recently brought to our attention.

6. Rowell, 1962.

7. Gaioni and Evans, 1985, 1986a, 1986b.

8. Morton and Page, 1992.

9. Lambrechts and Dhondt, 1986.

10. It may be that the conflict between the ability to listen and the ability to concentrate on talking is the foundation of lie detectors. The person taking a lie detector test may be on the other end of a phone line, under no fear of direct attack. What is it that prevents a liar's voice from sounding like that of a truthful person?

An inherent difference between a liar and one who is telling the truth is that the truthful person has a true story that does not need to be changed according to circumstance, while a liar is making up a story, and his or her success depends on the ability to convince the listener of something that isn't true. The liar has to expend effort to make up the story. And since the liar may not know in advance the listener's prior knowledge of the subject, he or she needs to pay close attention to the listener's reaction in order to adapt the story, change it slightly—and remember the changes—in order to be convincing. This close attention to the listener is very likely to affect muscles of the neck and head, and that in turn is likely to affect the voice to some degree. This difference may well be the one that lie detectors focus on (Streeter at al., 1977). If the liar tries to relax these muscles, his or her listening ability will decrease and with it the ability to lie successfully.

11. Anava, 1992.

12. Katsir, 1985, 1991, 1995.

13. Zahavi, 1978b.

14. Payne, 1983.

15. Hultsch and Todt, 1986; Todt and Hultsch, 1995.

16. Payne, 1983.

17. McGregor, 1993.

18. Pepperberg, 1991; Kaufman, 1991.

19. Ofer Hochberg, personal communication.

20. Alan Kemp, personal communication.

21. Hultsch and Todt, 1986.

22. Loffredo and Borgia, 1986.

23. Darwin, 1874.
24. Seyfarth et al., 1980.
25. Marler, 1955.
26. See chapter 12.

Chapter 7

1. Zuk et al., 1990.
2. Holder and Montgomerie, 1993.
3. Darwin, 1871.
4. Sutter, 1994.
5. Evans and Thomas, 1992.
6. Evans, 1991.
7. Andersson, 1982.
8. Eibel-Eibelsfeldt, 1970.
9. See also the discussion of fishes' fins in chapter 2.
10. Alex Kacelnick pointed out a logical weakness in our argument that manes and other features lessen the apparent size of body structures. After all, we assert that erect feathers and hair cannot be meant to increase apparent size since watchers can detect the deception. Yet we are suggesting an opposite deception—when we claim that a frame of bristling hair or feathers is a handicap that reduces the apparent size of the animal. Should watchers not disregard this deception as well, even if it is based on an optical illusion (Ponzo's effect—see Fujita et al., 1991)? We accept that the watcher may know that the mane decreases the apparent size of the shape within it. But this effect makes it more difficult for a slightly larger individual to show clearly its superiority in size, a superiority that would have been obvious if it were not for the mane. This is a real handicap: only an individual who is significantly larger than another can afford a decoration that decreases its apparent size without impairing its ability to show off that in fact it is larger than others.
11. Richard Wagner, personal communication.
12. Giora Ilani, personal communication.
13. See also the discussion of human beards in chapters 2 and 18.
14. Darling, 1937.
15. Clutton-Brock et al., 1982.
16. Moller, 1991.

Chapter 8

1. Hill, 1990.
2. Endler (1983, 1987), Lythgoe (1979), Hailman (1977), Butcher and Rohwer (1989), and others such as Hamilton, W.J. (1973), and Hingston (1933) have studied the benefits and drawbacks of specific colors. To them it was a question of balance between advertising over distance on the one hand and merging with the background to avoid enemies and predators on the other. They also studied the eye's sensitivity to various colors.
3. Mayr and Stresemann, 1950.
4. Diamond, 1987.

5. Maier, 1993.

6. Anderson, 1996.

Chapter 9

1. Eisner and Meinwald, 1987, 1995.

2. Schneider, 1992.

3. Ellis et al., 1980.

4. At the same time, though, high concentrations of female pheromone impair the ability of males to *find* females. In biological pest control, synthetic hormones are used to "confuse" males.

5. On intoxicating beverages and the value of showing off the ability to imbibe without getting drunk, see chapter 13.

6. Nahon et al., 1995.

7. Jackson and Hartwell, 1990a, 1990b.

8. Ulloa-Aguirre, 1995.

9. See the bibliography in Nahon et al., 1995.

10. See ibid.

11. Zahavi, 1993.

12. Snyder and Bredt, 1992.

13. See the bibliography in Nahon et al., 1995.

14. Zahavi, 1993.

Chapter 10

1. Zahavi, 1979.

2. Zahavi, 1971b.

3. Morris, 1956.

4. See chapter 12.

5. Selander, 1972.

6. See chapter 3.

7. Borgia, personal communication.

8. Osztreiher, 1992.

9. Spiro, 1963.

10. Van Lawick-Goodall, 1970.

11. Rasa, 1986.

Chapter 11

1. Trivers, 1974.

2. Trivers further complicated the problem by suggesting that according to kin selection, the child also has some interest in its parents' reproduction because of its genetic similarity to its siblings; see Trivers, 1972. This complication is unnecessary, as we shall see when we deal with the issue of kin selection in chapter 13.

3. Zahavi, 1977a.

4. Heinroth, 1926.

5. Feldman and Eshel, 1982.

6. See Redondo and Castro, 1992.
7. Mock, 1984.
8. Diamond, 1992.
9. Sade, 1972.

Chapter 12

1. See articles in Hebrew by Pozis (1984), Carmeli (1988), Katsir (1991), Osztreiher (1992), Anava (1992), Kalishov (1996), Perl (1996), Zahavi, T. (1975).
2. Later in this chapter we present graphs based on some of the data that have been collected so far, much of which is as yet available only in Hebrew.
3. Osztreiher, 1996.
4. Stachey and Koenig, 1990; Rowley and Russel, 1990.
5. Van Lawick-Goodall, 1971.
6. Kruuk, 1972.
7. Van Lawick-Goodall, 1970.
8. Rasa, 1986.
9. Sherman et al., 1991.
10. Bonner, 1967.
11. Rosenberg, 1984.
12. See Maynard Smith, 1964, 1976b.
13. Hamilton, 1964.
14. Trivers, 1971.
15. Axelrod, 1986.
16. Axelrod and Hamilton, 1981.
17. There is extensive literature on these models. See a detailed discussion and bibliography in Dawkins (1989) and a more up-to-date one in Sigmund (1993).
18. Stachey and Koenig, 1990.
19. See chapter 1, on prey–predator relations.
20. Zahavi, T., 1975.
21. Carlisle and Zahavi, 1986.
22. Kalishov, 1996.
23. Carlisle and Zahavi, 1986.
24. Kalishov, 1996.
25. Carmeli, 1988.
26. Slagsvold, 1984, 1985.
27. Carlisle and Zahavi, 1986; Zahavi, 1989.
28. Schjelderup-Ebbe, 1922; see Marler and Hamilton, 1966.
29. Perl, 1996.
30. Zahavi, 1988; Perl, 1996.
31. Perl, 1996.
32. Gaston, 1978. Only one species of babblers in India lives in pairs.
33. Komdeur, 1992, 1994.
34. Woolfenden and Fitzpatrick, 1984, 1990.
35. Koford et al., 1990.
36. Faaborg and Bednarz, 1990.
37. See Alexander, 1987; De Waal, 1996.

38. Zahavi, 1995.

39. Ibid.

Chapter 13

1. See a detailed discussion in Cronin, 1991.

2. Alexander, 1974.

3. See chapter 14.

4. Wilson, 1971.

5. West-Eberhard, in her study of *Polistes canadensis* (1986), showed that when one queen clearly controls the others, the level of aggression in the nest is low, while when there are only slight differences between the queens, aggression is high and there may be life-and-death struggles among the partners.

6. West, 1969; West-Eberhard, 1984; Gadagkar, 1991; Ito, 1993.

7. The wasp *Vespula germanica* was brought by Europeans to Australia. In tropical regions, it turned out that its colonies could survive the winter, and new colonies of this species in such regions are formed by coalitions of several queens. See Ito, 1993.

8. Wilson, 1971; Heinze et al., 1994.

9. Trivers, 1985.

10. It may be that some workers do indeed avoid working, but this does not demand any special explanation. It is the fact that workers *do* invest in the colony, as indeed most do, that demands explanation.

11. West, 1969.

12. Gadagkar, 1991; Heinze et al., 1994.

13. Marler and Hamilton (1966) define pheromones as chemicals secreted by one individual in order to elicit a specific reaction in another individual of the same species. As a rule, pheromones are mixtures of chemicals, and in most cases their specific components are not known. Thus, we use the terms *queen pheromone* or *pheromones* without getting into the specifics of one material or another.

14. Ishay et al. (1967, 1968) found that worker oriental hornets cannot make sugars out of proteins, a process known as *gluconeogenesis;* they feed the larvae proteins, and eat sugars they get from the larvae. If such mutual feeding is found between the larvae and the adult workers of other social insects, it will provide another reason—probably the main one—for workers to feed the larvae and take care of them. However, this very interesting area requires more research.

15. Winston and Slessor, 1992.

16. Engels and Imperatriz-Fonseca, 1990.

17. Van der Blom, 1986.

18. Roseler and Honk, 1989; see also Velthuis, 1990.

19. Diamond, 1990, 1992.

20. Veblen, 1899.

21. Trivers, 1974, 1985. Note that by this definition, a host taking care of its parasite's offspring is an altruist, and indeed Trivers so defines it (1985). By the same definition, when a helper is sterile to begin with, its reproduction cannot diminish any further and thus its help cannot be seen as altruism.

22. Trivers, 1971.

23. Hamilton, 1964.

24. Dawkins, 1989.

25. The calculation is: in sexually reproducing animals, a random half of one's genes come from the mother and the other half from the father. Sometimes the gene of one parent is not identical to but rather an allele of the gene of the other parent. Since the individual inherits randomly one or the other, there is a 50-percent probability that a given allele found in one individual will also be found in its sibling, and a 25-percent probability that they will also be found in the sibling's children (a probability of 50 percent times another probability of 50 percent).

26. A detailed discussion of this theory can be found in any book on sociobiology. See for example Trivers (1985); Dawkins (1989); Wilson (1975); West-Eberhard (1975); Krebs and Davies (1993).

27. Termite nymphs, like the larvae of ants, wasps, and bees, depend on their caretakers for food. Again, as in hymenoptera, termite workers are not completely sterile, and their fate depends on the treatment they get while growing. To understand what is happening in a termite colony, one would have to find out what it is that prevents individuals from striking out on their own, and how the workers' service to the colony benefits them individually.

28. Wilson, 1975; Krebs and Davies, 1993.

29. Gadagkar, 1991.

30. Hölldobler and Wilson, 1990, in their book about ants, tell of worker ants of the species *Myrmecocystis mumicus* and *Solenopsis invuca* who move to the biggest nest around and leave their mothers to die of hunger in the nest where they hatched. Such findings run contrary to the theory of kin selection but are in full accord with our suggestion that workers choose the colony where their chances of reproducing are best.

31. Haldane, 1932, 1955.

32. Motro and Eshel, 1988; Eshel and Motro, 1988.

33. In fact, some researchers who still accept group selection as a valid theory wonder why it is that most others in the profession accept kin selection as valid while rejecting group-selection models. See Wilson and Sober, 1994.

34. Zahavi, 1995.

35. Zahavi, 1974, 1989; Woolfenden and Fitzpatrick, 1990; Komdeur, 1994.

36. Gadagkar and Joshi, 1985.

37. Darwin, 1871.

38. Hölldobler and Wilson, 1990.

39. Dawkins, 1989.

40. West, 1969.

41. Roseler and Honk, 1989, remark that among bumblebees, the struggle for dominance among young queens starts in the first days following their emergence from the pupa stage.

42. Zahavi, 1995.

Chapter 14

1. Trivers, 1972, discussed in detail the implications of the conflict of interest between the genders, and many have studied the subject since. We are not attempting to challenge his treatment of the subject but rather to add some observations from our own perspective.

2. Selander, 1965, 1972; Orians, 1969; Emlen and Oring, 1977.

3. Gustafson, 1989; Beissinger, 1986; Beissinger and Snyder, 1987.

4. Dawkins and Carlisle, 1977.

5. Scott, 1988; Owen and Black, 1989; Forslund, 1990.

6. Newton, 1989; Gustafson, 1989.

7. See Gustafson, 1989.

8. Montgomerie, 1986; Zahavi, 1986.

9. Zahavi, 1988a; Perl, 1996.

10. Moller, 1990.

11. Morton et al., 1990; Wagner et al., 1996.

12. Zilberman, 1991.

13. Goldstein et al., 1986.

14. Sugiyama, 1967.

15. Timna, personal communication; Ilani, personal communication.

16. Trivers, 1972.

17. Reyer, 1990.

18. Munehara et al., 1994.

19. Kraak, 1994.

20. Fishman, 1977 (Hebrew).

21. Moreno et al., 1994.

22. Leader, 1996.

23. Orians and Beletsky, 1989.

24. Emlen and Oring, 1977; Emlen et al., 1989.

25. Gustafson, 1989.

Chapter 15

1. This description is based on Bonner (1991) and Nanjundiah and Saran (1992).

2. See Atzmony et al., 1997.

3. cAMP is a chemical with many roles in all living organisms—from bacteria to mammals, including humans. It is considered a secondary messenger transferring stimuli from the cell membrane to the inside of the cell.

4. Shaulski and Loomis (1993) found that some of the amebas that ended up as spores showed signs of having been prestalk before that. In other words, at the end of the process of migration and stalk formation, some of the prestalk amebas manage to become spores.

5. See Atzmony et al., 1997.

6. Bruce Levin, personal communication.

7. Atzmony and Nanjundiah, personal communication.

8. See Rosenberg, 1984.

9. See Zahavi and Ralt, 1984.

10. Shaulski and Loomis, 1995.

Chapter 16

1. Dawkins and Krebs, 1979; Rothstein, 1990.

2. Davies and Brook, 1989.

3. Lotem et al., 1991, 1995.

4. Lotem wrote another article about the dangers of evolving the ability to recognize a parasite's nestling and desert it or throw it out of the nest. See Lotem, 1993b.

5. Research done by Yoram Shpirer, Amotz Zahavi, Arnon Lotem, and Steve Rothstein.

6. Soler et al., 1995.

7. Zahavi, 1979.

8. See chapters 3, 12, and 14.

9. Newton, 1989b; Scott, 1988; Owen and Black, 1989.

10. Yom-Tov, 1989.

11. Ingle, 1911; Wyllie, 1975; Witherby et al., 1949.

12. Zahavi, 1979.

13. Soler et al., 1995.

14. J. Tengo, University of Uppsala, Sweden, personal communication.

15. Tengo, 1984.

16. M. Sorensen, personal communication.

17. Smith et al., 1984.

18. Oberski, 1975; Kuris, 1974.

19. Schwammberger, 1993.

20. Salyers and Whitt, 1994.

Chapter 17

1. Zahavi, 1971a.

2. Ward, 1965.

3. Darling, 1938.

4. Unfortunately, Ward, the partner with whom Amotz developed the idea of information centers, died at an early age.

5. Ward and Zahavi, 1973.

6. Parker-Rabenold, 1978; Broom et al., 1976; Heinrich, 1988.

7. Brown, 1986.

8. Heinrich, 1988.

9. Wagner, 1996.

10. Zahavi, 1983, 1995.

11. Ward, 1972.

12. Feare et al., 1974.

13. Heredia et al., 1991; Hiraldo et al., 1993.

14. Bucher, 1992.

15. Wynne-Edwards, 1962.

16. Marx, personal communication.

17. Veblen, 1899.

18. Ibid.

19. See Wynne-Edwards, 1962.

20. In our original article on information centers (Ward and Zahavi, 1973) we made something of a group-selection argument ourselves. We have since recognized that early error and have suggested that other explanations should be sought for the communal displays at roosts (Zahavi, 1985b). This subject has received further treatment in our response to an article by Richner and Heeb (Zahavi, 1996).

21. Zahavi, 1983, 1995.

22. See chapter 1.

Chapter 18

1. Shepher, 1983.
2. Hess, 1965.
3. Eibel-Eibelsfeldt, 1971.
4. See chapter 8.
5. 2 Samuel 20:9.
6. See chapter 7.
7. Morris, 1967.
8. Caro and Seller, 1990.
9. Sever and Mendelsohn, 1989.
10. De Waal, 1995.
11. Strahl, 1988.
12. See chapter 10.
13. Zahavi, 1971b.
14. Trivers, 1985.
15. Wagner, 1996.
16. See chapter 11.
17. The gray parrot Alex; see Pepperberg, 1991.
18. The exception is among the most highly social bees, wasps, ants, and termites, whose communities include thousands of individuals.
19. The differences between a language of nonverbal vocalizations and a verbal language is like the difference between an analog speedometer and a digital speedometer. An analog speedometer's needle lets us estimate speed pretty precisely, even if there are only two or three numbers on the face of the speedometer. A digital speedometer, which displays numbers, is limited to the precision of these numbers. If the digits change with only every 5 additional miles per hour, the digital speedometer will not show the difference between, say, 11 mph and 14 mph—while an analog speedometer will.
20. Zahavi, 1980.
21. Frith and Frith, 1990; Diamond, 1991.
22. See chapter 4.
23. Benedict, 1946.
24. Bakal, 1979.
25. Veblen, 1899.

BIBLIOGRAPHY

Alcock, J. 1975. 5th ed. 1993. *Animal behaviour: An evolutionary approach.* Sunderland, MA: Sinauer Associates.

Alexander, R.D. 1974. The evolution of social behaviour. *Ann. Rev. Ecol. Syst.* 5: 325–83.

———. 1987. *The Biology of Moral Systems.* New York: Aldine De Gruyter.

Anava, A. 1992. The value of mobbing behaviour for the individual babbler (*Turdoides squamiceps*) (in Hebrew). M.S. thesis, Ben-Gurion University of the Negev.

Anderson, S. 1996. Bright UV colour in the Asian whistling thrushes. *Proc. R. Soc. Lond. B.,* 263, 843–48.

Andersson, M. 1982. Female choice selects for extreme tail length in widow birds. *Nature* 299: 818–20.

———. 1994. *Sexual selection.* Princeton: Princeton University Press.

Atzmony, D., A. Zahavi and V. Nanjundiah. 1997. Altruistic behaviour in *Dictyostelium discoideum* explained on the basis of individual selection. *Current Science* 72: 142-45.

Axelrod, R. 1986. An evolutionary approach to norms. *Amer. Politic. Sci. Rev.* 80: 1095–1111.

Axelrod, R., and W.D. Hamilton. 1981. The evolution of cooperation. *Science* 211: 1390–96.

Baerends, G.P., and J.M. Baerends-van Roon. 1950. An introduction to the study of the ethology of cichlid fishes. *Behaviour* (suppl.) 1: 1–242.

Bakal, C. 1979. *Charity U.S.A.: An investigation into the hidden world of the multi-billion dollar charity industry.* New York: Times Books.

Barlow, J.W. 1972. The attitude of fish eye-lines in relation to body shape and to stripes and bars. *Copeia* 72: 4–12.

Barrete, C., and D. Vandal. 1990. Sparring, relative antler size, and assessment in male caribou. *Behav. Ecol. Sociobiol.* 26: 383–87.

Basolo, L.A. 1990. Female preference predates the evolution of the sword in swordfish. *Science* 250: 808–10.

Beissinger, S.R. 1986. Demography, environmental uncertainty, and the evolution of mate desertion in the snail kite. *Ecology* 67: 1445–59.

Beissinger, S.R., and N.F.R. Snyder. 1987. Mate desertion in the snail kite. *Anim. Behav.* 35: 477–87.

Benedict, R. 1946. *Patterns of culture.* New York: Penguin Books.

Boake, C.R.B. 1991. Coevolution of senders and receivers of sexual signals: genetic coupling and genetic correlations. *Tree* 6: 225–27.

Bonner, J.T. 1967. *The cellular slime molds.* Princeton: Princeton University Press.

Borgia, G. 1986. Sexual selection in bowerbirds. *Sci. Amer.* 254: 70–79.

Borgia, G., and K. Collins. 1986. Feather stealing in the satin bowerbird (*Ptilonorhynchus violaceus*): Male competition and the quality of display. *Anim. Behav.* 34: 727–38.

Broom, D.M., W.J.A. Dick, C.E. Johnson, D.I. Sales, and A. Zahavi. 1976. Pied wagtail roosting and feeding behaviour. *Bird Study* 23: 267–80.

Brown, C.R. 1986. Cliff swallow colonies as information centers. *Science* 234: 83–85.

Bucher, E.H. 1992. The causes of extinction of the passenger pigeon. In *Current Ornithology,* ed. D.N. Power, vol. 9, 1–26. New York: Plenum Press.

Burley, N. 1986. Sexual selection for aesthetic traits in species with biparental care. *Amer. Nat.* 127: 415–45.

Burley, N., G. Krantzberg, and P. Radman. 1982. Influence of colour-banding on the conspecific preferences of zebra finches. *Anim. Behav.* 30: 444–55.

Buss, L.W. 1987. *The evolution of individuality.* Princeton: Princeton University Press.

Butcher, G.S., and S. Rohwer. 1989. The evolution of conspicuous and distinctive coloration for communication in birds. *Current Ornithol.* 6: 51–108.

Carlisle, T.R., and A. Zahavi. 1986. Helping at the nest, allofeeding and social status in immature Arabian babblers. *Behav. Ecol. Sociobiol.* 18: 339–51.

Carmeli, Z. 1988. Mobbing behaviour in the Arabian babbler (*Turdoides squamiceps*) (in Hebrew). M.S. thesis, Hebrew University, Jerusalem.

Caro, T.M. 1994. Ungulate antipredator behaviour: Preliminary and comparative data from African bovids. *Behaviour* 128: 189–228.

Caro, T.M., and D.W. Seller. 1990. The reproductive advantages of fat in women. *Ecol. and Sociobiol.* 11: 51–66.

Christy, J.H. 1988. Pillar function in the fiddler crab *Uca beebei.* II: Competitive courtship signalling. *Ethology* 78: 113–28.

Clutton-Brock, T.H., and S.D. Albon. 1978. The roaring of red-deer and the evolution of honest advertising. *Behaviour* 69: 143–69.

Clutton-Brock, T.H., F.E. Guinness, and S.D. Albon. 1982. *Red Deer: The Behaviour and Ecology of Two Sexes.* Chicago: University of Chicago Press.

Collins, S. 1993. Is there only one type of male handicap? *Proc. R. Soc. Lond. B.* 252: 193–97.

Cresswell, W. 1994. Song as a pursuit-deterrent signal, and its occurrence relative to other anti-predation behaviours of skylark *(Alauda arvensis)* on attack by merlins *(Falco columbarius). Behav. Ecol. Sociobiol.* 34: 217–23.

Cronin, H. 1991. *The ant and the peacock.* Cambridge, UK: Cambridge University Press.

Cullen, J.M. 1966. Reduction of ambiguity through ritualization. *Phil. Trans., R. Soc. B.* 251: 363–74.

Curio, E. 1978. The adaptive significance of avian mobbing. *Z. Tierpsychol.* 48: 175–83.

Darling, F.F. 1937. *A Herd of Red Deer.* Oxford: Oxford University Press.

———. 1938. *Bird flocks and the breeding cycle.* Cambridge: Cambridge University Press.

Darwin, C. 1859. *On the origin of species by means of natural selection, or the preservation of favoured races in the struggle for life.* London: John Murray.

———. 1871. *The descent of man and selection in relation to sex.* Facsimile ed. Princeton: Princeton University Press, 1981.

———. 1872. *The expression of the emotions in man and animals.* London: John Murray.

Davies, N.B., and M. Brook. 1989. An experimental study of co-evolution between the cuckoo *Cuculus canorus* and its hosts. II: Host egg markings, chick discrimination and general discussion. *J. Anim. Ecol.* 58: 225–36.

Davies, N.B., and T.R. Halliday. 1978. Deep croaks and fighting assessment in toads *(Bufo bufo). Nature* 274: 683–85.

Davis, G.W.F., and P. O'Donald. 1976. Sexual selection for a handicap: A critical analysis of Zahavi's model. *J. Theor. Biol.* 57: 345–54.

Dawkins, R. 1989. *The selfish gene.* 2nd ed. Oxford: Oxford University Press.

Dawkins, R., and T.R. Carlisle. 1976. Parental investment, mate desertion and a fallacy. *Nature* 262: 131–33.

Dawkins, R., and J.R. Krebs. 1979. Arms races between and within species. *Proc. Roy. Soc. Lond. B.* 205: 489–511.

De Waal, F.B.M. 1995. Bonobo sex and society. *Sci. Am.* 272 (3): 58–65.

De Waal, F.B.M. 1996. *Good natured: The origins of right and wrong in humans and other animals.* Cambridge, MA: Harvard University Press.

Diamond, J. 1986a. Animal art: Variation in bower decorating style among male bowerbirds *(Amblyornis inornatus). Proc. Natl. Acad. Sci. USA.* 83: 3042–46.

———. 1986b. Biology of birds of paradise and bowerbirds. *Ann. Rev. Ecol. Sys.* 17: 17–37.

———. 1987. Flocks of brown and black new-guinean birds: A bicoloured mixed-species foraging association. *Emu* 87: 201–11.

———. 1990. Kung fu kerosene drinking. *Natural History* 7: 20–24.

———. 1992. *The rise and fall of the third chimpanzee.* London: Radius.

Diamond, J.M. 1991. Borrowed sexual ornaments. *Nature* 349: 105.

Dominey, W.J. 1980. Female mimicry in male bluegill sunfish—a genetic polymorphism? *Nature* 284: 546–48.

Eibel-Eibelsfeldt, I. 1961. The fighting behavior of animals. *Sci. Amer.* 205: 112–28.

———. 1970. *Ethology, the Biology of Behavior.* New York: Holt, Rinehart, and Winston.

———. 1971. *Love and Hate.* London: Methuen.

Eisner, T., and J. Meinwald. 1987. Alkaloid-derived pheromones and sexual selection in Lepidoptera. In *Pheromone biochemistry,* ed. G.D. Prestwich and G.J. Blomquist, 251–69. Orlando, FL: Academic Press.

————. 1995. The chemistry of sexual selection. *Proc. Natl. Acad. Sci. USA* 92: 50–55.

Ellis, P.E., L.C. Brimacombe, L.J. McVeigh, and A. Dignan. 1980. Laboratory experiments on the disruption of mating in the Egyptian cotton leafworm *Spodoptera littoralis* (Lepidoptera: Noctuidae) by excesses of female pheromones. *Bull. Ent. Res.* 70: 673–84.

Emlen, S.T., N.J. Demong, and D.J. Emlen. 1989. Experimental induction of infanticide in female wattled jacanas. *Auk* 106: 1–7.

Emlen, S.T., and L.W. Oring. 1977. Ecology, sexual selection and the evolution of mating systems. *Science* 197: 215–23.

Endler, J.A. 1983. Natural and sexual selection of color patterns in Poeciliid fishes. *Environ. Biol. of Fishes* 9: 173–90.

————. 1987. Predation, light intensity, and courtship behaviour in *Poecilia reticulata* (Pisces: Poeciliidae). *Anim. Behav.* 35: 1376–85.

Engels, W., and V.L. Imperatriz-Fonseca. 1990. Caste development, reproductive strategies and control of fertility in honey bees and stingless bees. In *Social insects: An evolutionary approach to caste and reproduction,* ed. W. Engels, 167–230. Berlin: Springer-Verlag.

Eshel, I., 1978a. A critical defence of the handicap principle. *J. Theor. Biol.* 70: 245–50.

————. 1978b. On a prey-predator nonzero-sum game and the evolution of gregarious behavior of evasive prey. *Amer. Nat.* 112: 787–95.

Eshel, I., and U. Motro. 1988. The three brothers' problem—kin selection with more than one potential helper. I: The case of immediate help. *Amer. Nat.* 132: 550–66.

Evans, C.S. 1985. Display vigour and subsequent fight performance in the Siamese fighting fish *Betta splendens*. *Behavioural Processes* 11: 113–21.

Evans, M.R. 1991. The size of adornments of male scarlet-tufted malachite sunbirds varies with environmental conditions, as predicted by handicap theories. *Anim. Behav.* 42: 797–803.

Evans, M.R., and A.L.R. Thomas. 1992. The aerodynamic and mechanical effects of elongated tails in the scarlet-tufted malachite sunbird: the cost of a handicap. *Anim. Behav.* 43: 337–47.

Ewer, R.F. 1968. *Ethology of mammals*. London: Logos Press.

Faaborg, J., and J.C. Bednarz. 1990. Galapagos and Harris' hawks: Divergent causes of sociality in two raptors. In *Cooperative breeding in birds,* ed. P.B. Stacey and W.D. Koenig, 357–84. New York: Cambridge University Press.

Feare, C.J., G.M. Dunnet, and I.J. Patterson. 1974. Ecological studies of the rook (*Corvus frugilegus* L.) in north-east Scotland; food intake and feeding behavior. *J. Appl. Ecol.* 11: 867–896.

Feldman, M.W., and I. Eshel. 1982. On the theory of parent-offspring conflict: A two-locus genetic model. *Amer. Nat.* 119: 285–92.

Fisher, R.A. 1930. *The genetical theory of natural selection*. London: Clarendon Press.

Fishman, L. 1977. Wheatears—live warning systems against snakes (in Hebrew). *Teva Va'aretz* 19: 198.

FitzGibbon, C.D., and J.H. Fanshawe. 1988. Stotting in Thompson's gazelle: An honest signal of condition. *Behav. Ecol. Sociobiol.* 23: 69–74.

Forslund, P. 1990. Mate change reduces reproduction in the barnacle goose (*Branta leucopsis*)—the advantage of having an old partner. *Acta XX Con. Inter. Ornithol.* 470.

Foster, M.S. 1981. Cooperative behavior and social organization in the swallow-tailed manakin *(Chiroxiphia caudata). Behav. Ecol. Sociobiol.* 9: 167–77.

Frith, C.B., and D.W. Frith. 1990. Archbold's bowerbird, *Archboldia papuensis* (Ptilonorhynchidae), uses plumes from king bird of paradise, *Pteridophora alberti* (Paradisaeidae), as bower decoration. *Emu* 90: 136–37.

Fugle, G.N., S.I. Rothstein, C.W. Osenberg, and M.A. McFinley. 1984. Signals of status in wintering white-crowned sparrows *(Zonotrichia leucoptrys gambeli). Anim. Behav.* 32: 86–93.

Fujita, K., D.S. Blough, and P.M. Blough. 1991. Pigeons see the Ponzo illusion. *Anim. Learning Behav.* 19: 283–93.

Gadagkar, R. 1991. Belonogaster, Mischocyttarus, Parapolybia, and independent-founding Ropalidia. In *The social biology of wasps,* ed. K.G. Ross and R.W. Matthews, 149–200. Ithaca, NY: Cornell University Press.

Gadagkar, R., and N.V. Joshi. 1985. Colony fission in a social wasp. *Current Sci.* 54: 57–62.

Gaioni, S.J., and C.S. Evans. 1985. The role of frequency modulation in controlling the response of mallard ducklings *(Anas platyrhynchos)* to conspecific distress calls. *Anim. Behav.* 33: 188–200.

———. 1986a. Mallard duckling response to distress calls with reduced variability: Constraint on stereotypy in a "fixed action pattern." *Ethology* 72: 1–14.

———. 1986b. Perception of distress calls in mallard ducklings *(Anas platyrhynchos). Behaviour* 99: 250–74.

Gaston, A.J. 1977. Social behaviour within groups of jungle babblers *(Turdoides striatus). Anim. Behav.* 25: 828–48.

———. 1978. The evolution of group territorial behaviour and cooperative breeding. *Amer. Nat.* 112: 1091–1100.

Gibbs, H.L., and P.R. Grant. 1987. Oscillating selection on Darwin's finches. *Nature* 327: 511–13.

Gibson, R.M., J.W. Bradbury, and S.L. Vehrencamp. 1991. Mate choice in lekking sage grouse revisited: The roles of vocal display, female site fidelity and copying. *Behav. Ecol.* 2: 165–80.

Gibson, R.M., and J. Hoglund. 1992. Copying and sexual selection. *Tree* 7: 229–32.

Goldstein, H., D. Eisikovitz, and Y. Yom-Tov. 1986. Infanticide in the Palestine sunbirds *Nectarina osea. Condor* 88: 528–29.

Grafen, A. 1990a. Biological signals as handicaps. *J. Theor. Biol.* 144: 517–46.

———. 1990b. Sexual selection unhandicapped by the Fisher process. *J. Theor. Biol.* 144: 473–516.

Greenewalt, C.H. 1986. *Bird song: Acoustics and physiology.* Washington, DC: Smithsonian Institution Press.

Gustafsson, L. 1989. Collared flycatcher. In *Lifetime reproduction in birds,* ed. I. Newton, 75–88. London: Academic Press.

Hailman, J.P. 1977. *Optical signals: Animal communication and light.* Bloomington, IN: Indiana University Press.

Haldane, J.B.S. 1932. *The Causes of Evolution.* London: Longmans Green. Reprint. Princeton: Princeton Science Library, 1990.

———. 1955. Population genetics. *New Biology* 18: 34–51.

Hamilton, W.D. 1964. The genetical evolution of social behaviour. *J. Theor. Biol* 7: 1–52.

Hamilton, W.D., and M. Zuk. 1982. Heritable true fitness and bright birds: A role for parasites? *Science* 218: 384–87.

Hamilton, W.J. III. 1973. *Life's color code.* New York: McGraw-Hill.

Hunter, C.P., and P.D. Dwyer. The value of objects to satin bowerbirds (Ptilonorhynchus violaceus). *Emu* (in press).

Hasson, O. 1991a. Pursuit-deterrent signals: Communication between prey and predator. *Tree* 6: 325–29.

———. 1991b. Sexual displays as amplifiers: Practical examples with an emphasis on feather decorations. *Behav. Ecol.* 2: 189–97.

Hasson, O., R. Hibbard, and G. Cebballos. 1989. The pursuit-deterrent function of tail wagging in the zebra-tailed lizard *(Callisaurus draconoides). Canad. J. Zool.* 67: 1203–09.

Heinrich, B. 1988. Winter foraging at carcasses by the three sympatric corvids, with emphasis on recruitment by the raven, *Corvus corax. Behav. Ecol. Sociobiol.* 23: 141–56.

Heinroth, O.J., and M. Heinroth. 1926. *Die Vögel Mitteleuropas.* Berlin-Lichterfelde: Hugo Bermuhler-Verlag.

Heinze, J., B. Hölldobler, and C. Peeters. 1994. Conflict and cooperation in ant societies. *Naturwissenschaften* 81: 489–97.

Hendry, L.B., E.D. Bransome, M.S. Hutson, and L.K. Campbell. 1984. A newly discovered stereochemical logic in the structure of DNA suggests that the genetic code is inevitable. *Perspectives in Biology and Medicine* 27: 623–51.

Hendry, L.B., and V.B. Mahesh. 1995. A putative step in steroid hormone action involves insertion of steroid ligands into DNA facilitated by receptor proteins. *J. Steroid Biochem. Molec. Biol.* 55: 173–83.

Heredia, B., J.C. Alonso, and F. Hiraldo. 1991. Space and habitat use by red kites, *Milvus milvus,* during winter in the Guadalquivir marshes: A comparison between resident and wintering populations. *Ibis* 133: 374–81.

Hess, H.E. 1965. Attitude and pupil size. *Sci. Amer.* 212: 46–54.

Hill, G.E. 1990. Female house finches prefer colorful mates: Sexual selection for a condition-dependent trait. *Anim. Behav.* 40: 563–72.

Hingston, R.W.G. 1933. *The meaning of animal colour and adornment.* London: Arnold.

Hiraldo, F., B. Heredia, and J.C. Alonso. 1993. Communal roosting of wintering red kites, *Milvus milvus* (Aves, Accipitridae): Social feeding strategies for the exploitation of food resources. *Ethology* 93: 117–24.

Hoglund, J., R. Montgomerie, and F. Widemo. 1993. Costs and consequences of variation in the size of ruff leks. *Behav. Ecol. Sociobiol.* 32: 31–39.

Hogan-Warburg, A.J. 1966. Social behavior of the ruff, *Philomachus pugnax* (L). *Ardea* 54: 109–229.

Holder, K. and R. Montgomerie. 1993. Context and consequences of comb displays by male rock ptarmigan. *Anim. Behav.* 45: 457–70.

Hölldobler, B., and E.O. Wilson. 1990. *The ants.* Cambridge: Harvard University Press.

Hultsch, H., and D. Todt. 1986. Signal matching: Zeichenbildung durch mustergleiches Antworten. *Semiotik* 8: 233–44.

Huxley, J.S. 1914. The courtship habits of the great crested grebe *(Podiceps cristatus)* with an addition to the theory of sexual selection. *Proc. Zool. Soc. London* 35: 491–562.

Ingle, A. 1911. Cuckoos as nest robbers. *Emu* 11: 254-55.

Ishay, J., H. Bitinsky-Salz, and A. Shulov. 1967. Contributions to the bionomics of the oriental hornet (*Vespa orientalis* Fab.). *Israel J. Entomol.* 2: 45–106.

Ishay, J., and R. Ikan. 1968. Food exchange between adults and larvae in *Vespa orientalis*. *Anim. Behav.* 16: 298–303.

Ito, Y. 1993. *Behaviour and social evolution of wasps.* Oxford Series in Ecology and Evolution. Oxford: Oxford University Press.

Jackson, C.L., and L.H. Hartwell. 1990a. Courtship in *S. cerevisiae:* Both cell types choose mating partners by responding to the strongest pheromone signal. *Cell* 63: 1039–51.

———. 1990b. Courtship in *Saccharomyces cerevisiae:* An early cell-cell interaction during mating. *Cell Biol.* 10: 2202–13.

Jarvi, T., and M. Baker. 1984. The function of the variation in the breast stripe of the great tit (*Parus major*). *Anim. Behav.* 32: 590–96.

Jones, I.L., and F.M. Hunter. 1993. Mutual sexual selection in a monogamous seabird. *Nature* 362: 238.

Kalishov, A. 1996. Allofeeding among babblers (*Turdoides squamiceps*) (in Hebrew, with English summary). M.S. thesis, Tel-Aviv University.

Katsir, Z. 1985. Vocal signal affected by body posture and movement in the Arabian babbler (*Turdoides squamiceps,* Aves: Timaliidae). *19th International Ethological Conference,* 2: 398. Toulouse, France: Sabatier University.

———. 1991. Messages in vocal communication: Investigations of the variations in the babbler "shout" (in Hebrew). Ph.D. thesis, Hebrew University, Jerusalem.

Katsir, Z. 1995. The meaning of the "variations" in the babbler "shout": A musical-ethological approach. *Behav. Processes* 34: 213–32.

Katzir, G. 1981a. Aggression by the damselfish, *Dascyllus aruanus* L., towards conspecifics and heterospecifics. *Anim. Behav.* 29: 835–41.

———. 1981b. Visual aspects of species recognition in the damselfish, *Dascyllus aruanus*. *Anim. Behav.* 29: 842–49.

Kaufman, K. 1991. The subject is Alex. *Audubon,* September–October, 52–58.

Kirkpatrick, M. 1986. The handicap mechanism of sexual selection does not work. *Am. Nat.* 127: 222–40.

Koford, R.R., B.S. Bowen, and S.L. Vehrencamp. 1990. Groove-billed anis: Joint-nesting in a tropical cuckoo. In *Cooperative breeding in birds,* ed. P.B. Stacey and W.D. Koenig, 333–56. New York: Cambridge University Press.

Komdeur, J. 1992. Importance of habitat saturation and territory quality for evolution of cooperative breeding in the Seychelles warbler. *Nature* 358: 493–95.

———. 1994. Experimental evidence for helping and hindering by previous offspring in the cooperative-breeding Seychelles warbler, *Acrocephalus seychellensis*. *Behav. Ecol. Sociobiol.* 34: 175–86.

Kraak, S.B.M. 1994. Female mate choice in *Aidablennius sphynx,* a fish with parental care for eggs in a nest. Ph.D. thesis, University of Groningen.

Krebs, J.R., and R. Dawkins. 1984. Animal signals: Mind-reading and manipulation. In *Behavioral ecology: An evolutionary approach,* ed. J.R. Krebs and N.B. Davies, 2nd ed., pp. 380–402. Oxford: Blackwell Scientific Publications.

Krebs, J.R., and N.B. Davies. 1993. *An introduction to behavioural ecology.* Oxford: Blackwell Scientific Publications.

Kruuk, H. 1972. *The spotted hyena*. Chicago: University of Chicago Press.

Kuris, A.M. 1974. Trophic interaction: Similarity of parasitic castrators to parasitoids. *Q. Rev. Biol.* 49: 129–48.

Lack, D. 1968. *Ecological adaptations for breeding in birds*. London: Chapman and Hall.

———. 1970. The endemic ducks of remote islands. *Wildfowl* 21: 5–10.

Lambrechts, M., and A. Dhondt. 1986. Male quality, reproduction, and survival in the great tit *(Parus major)*. *Behav. Ecol. Sociobiol.* 19: 57–63.

Leader, N., 1996. The function of stone ramparts at the entrance of blackstarts' *(Cercomela melanura)* nests (in Hebrew). M. Sc. thesis, Tel-Aviv University.

Loffredo, C.A., and G. Borgia. 1986. Male courtship vocalizations as cues for mate choice in satin bowerbird *(Ptilonorhynchus violaceus)*. *Auk* 109: 189–95.

Lorenz, K. 1952. *King solomon's ring*. London: Methuen.

———. 1962. The function of colour in coral reef fishes. *Proc. R. Inst. Gr. Br.* 39: 282–96.

———. 1966. *On aggression*. London: Methuen.

Lotem, A. 1993a. Secondary sexual ornaments as signals: The handicap approach and three potential problems. *Etologia* 3: 209–18.

———. 1993b. Learning to recognize nestlings is maladaptive for cuckoo *(Cuculus canorus)* hosts. *Nature* 362: 743–45.

Lotem, A., H. Nakamura, and A. Zahavi. 1991. Rejection of cuckoo egg in relation to host age: A possible evolutionary equilibrium. *Behav. Ecol.* 3: 128–32.

———. 1995. Constraints on egg discrimination and cuckoo-host co-evolution. *Anim. Behav.* 11: 43–54.

Lythgoe, J.N. 1979. *The ecology of vision*. Oxford: Clarendon Press.

Maier, E.J. 1993. To deal with the "invisible": On the biological significance of ultraviolet sensitivity in birds. *Naturwissenschaften* 80: 476–78.

Marler, P. 1955. Characteristics of some animal calls. *Nature* 176: 6.

Marler, P., and W.J. Hamilton. 1966. *Mechanisms of animal behavior*. New York: John Wiley and Sons.

Mathews, C.K., E.M. Kutter, G. Mosig, and P.B. Berget. 1983. Bacteriophage T4. *American Soc. for Microbiology,* Washington, D.C.

Maynard Smith, J. 1964. Group selection and kin selection. *Nature* 201: 1145–47.

———. 1965. The evolution of alarm calls. *Amer. Nat.* 99: 183–88.

———. 1976a. Evolution and the theory of game. *Am. Sci.* 64: 41–45.

———. 1976b. Sexual selection and the handicap principle. *J. Theor. Biol.* 57: 239–42.

———. 1976c. Group selection. *Quar. Rev. Biol.* 51: 277–83.

———. 1982. Do animals convey information about their intentions? *J. Theor. Biol.* 97: 1–5.

———. 1991a. Theories of sexual selection. *Tree* 6: 146–51.

———. 1991b. Must reliable signals always be costly? *Anim. Behav.* 47: 1115–20.

Maynard Smith, J., and G.A. Parker. 1976. The logic of asymmetric contests. *Anim. Behav.* 24: 159–75.

Mayr, E. 1942. *Systematics and the origin of species*. New York: Columbia University Press.

———. 1972. Sexual selection and natural selection. In *Sexual selection and the descent of man 1871–1971,* ed. B. Campbell, 97. Los Angeles: University of California Press.

Mayr, E., and E. Stresemann. 1950. Polymorphosm in the chat genus *Oenanthe* (Aves). *Evolution* 4: 291–300.

McDonald, D.B., and W.K. Potts. 1994. Cooperative display and relatedness among males in a lek-mating bird. *Science* 266: 1030–32.

McGregor, P.K. 1993. Signalling in territorial systems: A context for individual identification, ranging and eavesdropping. *Phil. Trans. R. Soc. Lond.* B 340: 237–44.

McKaye, K.R. 1991. Sexual selection and the evolution of the cichlid fishes of Lake Malawi. In *Cichlid fishes' behaviour, ecology and evolution,* ed. M.H.A. Keenleyside, 241–57. London: Chapman and Hall.

Mock, D.W. 1984. Siblicidal aggression and resource monopolization in birds. *Science* 225: 731–33.

Moller, A.P. 1990a. Fluctuating asymmetry in male sexual ornaments may reliably reveal male quality. *Anim. Behav.* 40: 1185–87.

———. 1990b. Male tail length and female mate choice in the monogamous swallow, *Hirundo rustica. Anim. Behav.* 39: 458–65.

———. 1990c. Sexual behavior is related to badge size in the house sparrow *Passer domesticus. Behav. Ecol. Sociobiol.* 27: 23–29.

———. 1991. Influence of wing and morphology on the duration of song flights in skylarks. *Behav. Ecol. Sociobiol.* 28: 309–14.

———. 1992. Female swallow preferences for symmetrical male sexual ornaments. *Nature* 357: 238–240.

———. 1994. *Sexual selection and the barn-swallow.* Oxford: Oxford University Press.

Montgomerie, R. (convenor). 1986. Symposium on mate guarding. *Acta XIX Con. Inter. Ornithol.* 408–53. Ottawa.

Moore, A.J. 1988. Female preferences, male social status, and sexual selection in *Nauphoeta cinerea. Anim. Behav.* 36: 303–5.

Moreno, J., M. Soler, A.P. Moller, and M. Linder. 1994. The function of stone carrying in the black wheatear *(Oenanthe leucora). Animal Behav.* 47: 1297–1309.

Morris, D. 1956. The function and causation of courtship ceremonies. In *L'instinct dans le comportement des animaux et de l'homme,* ed. P.P. Grasse, 261–86. Fondation Singer-Polignae. Paris: Masson.

———. 1957. "Typical intensity" and its relationship to the problem of ritualization. *Behaviour* 11: 1–12.

———. 1967. *The naked ape.* New York: McGraw-Hill.

———. 1990. *Animalwatching.* New York: Crown.

Morton, E.S. 1977. On the occurrence and significance of motivation-structural rules in some birds and mammal sound. *Amer. Nat.* 111: 855–69.

Morton, E.S., L. Forman, and M. Braun. 1990. Extrapair fertilizations and the evolution of colonial breeding in purple martins. *Auk* 107: 275–83.

Morton, E.S., and J. Page. 1992. *Animal talk.* New York: Random House.

Motro, U., and I. Eshel. 1988. The three brothers' problem: Kin selection with more than one potential helper. II: The case of delayed help. *Amer. Nat.* 132: 567–75.

Munehara, H., A. Takenaka, and O. Takenaka. 1994. Alloparental care in the marine sculpin *(Alcichthys alcicornis,* Pisces: Cottidae): Copulating in conjunction with parental care. *J. Ethol.* 12: 115–20.

Murray, I.R., and J.L. Arnott. 1993. Towards the simulation of emotion in synthetic speech: A review of the literature on human vocal emotion. *J. Acoust. Soc. Am.* 92: 1097–1108.

Nahon, E., D. Atzmony, A. Zahavi, and D. Granot. 1995. Mate selection in yeast: A reconsideration of the signals and the message encoded by them. *J. Theor. Biol.* 172: 315–22.

Nanjundiah, V., and S. Saran. 1992. The determination of spatial pattern in *Dictyostelium discoideum. J. Biosci.* 17: 353–93.

Newton, I. 1989a. Sparrowhawk. In *Lifetime reproduction in birds,* ed. I. Newton, 279–96. London: Academic Press.

———. 1989b. Synthesis. In *Lifetime reproduction in birds,* ed. I. Newton, 279–96. London: Academic Press.

Nisbet, I.C.T. 1973. Courtship feeding, egg size and breeding success in common terns. *Nature* 241: 141–42.

———. 1977. Courtship feeding and clutch size in common terns (*Sterna hirundo*). In *Evolutionary ecology,* ed. B. Stonehouse and C.M. Perrins, 101–9. London: Macmillan.

Norris, K.J. 1990. Female choice and the evolution of conspicious plumage coloration of monogamous male great tits. *Behav. Ecol. Sociobiol.* 26: 129–38.

Oberski, S. 1975. Parasite reproduction strategy and the evolution of castration of hosts by parasites. *Science* 188: 1314–16.

O'Donald, P. 1963. Sexual selection and territorial behaviour. *Heredity* 18: 361–64.

Orians, G. 1969. On the evolution of mating systems in birds and mammals. *Amer. Nat.* 103: 589–603.

Orians, G.H., and L.D. Beletsky. 1989. Red-winged blackbird. In *Lifetime reproduction in birds,* ed. I. Newton, 183–98. London: Academic Press.

Osztreiher, R. 1992. The morning dance of the Arabian babbler, *Turdoides squamiceps* (in Hebrew). M.S. thesis, Tel-Aviv University.

———. 1996. The competitive relationship among Arabian babbler (*Turdoides squamiceps*) nestlings (in Hebrew). Ph.D. thesis, Tel-Aviv University.

Owen, M., and J.M. Black. 1989. Barnacle goose. In *Lifetime reproduction in birds,* ed. I. Newton, 349–62. London: Academic Press.

Parker-Rabenold, P. 1987. Recruitment to food in black vultures: Evidence for following from communal roosts. *Anim. Behav.* 35: 1775–85.

Parsons, P.A. 1990. Fluctuating asymmetry: An epigenetic measure of stress. *Biol. Rev. Cambr. Phil. Soc.* 65: 131–45.

Payne, R.B. 1983. The social context of song mimicry: Song-matching dialects in indigo bunting (*Passerina cyanea*). *Anim. Behav.* 31: 788–805.

Pepperberg, I.M. 1991. A communicative approach to animal cognition: A study of conceptual abilities of an African grey parrot. In *Cognitive ethology: The minds of other animals,* ed. C.A. Ristau, 153–86. Hillsdale, NJ, and London: Lawrence Erlbaum Associates.

Perl, J. 1996. Competition for breeding between Arabian babbler males (in Hebrew). M.S. thesis, Tel-Aviv University.

Petrie, M., T. Halliday, and C. Sanders. 1991. Peahens prefer peacocks with elaborate trains. *Anim. Behav.* 41: 323–31.

Pliske, T.E., and T. Eisner. 1969. Sex pheromone of the queen butterfly: Biology. *Science* 164: 1170–72.

Pomiankowski, A. 1987. Sexual selection: The handicap principle does work—sometimes. *Proc. R. Soc. Lond. B.* 231: 123–45.

Pond, G. 1973. *An introduction to lace.* London: Garnstone Press.

Pozis, O. 1984. Play in babblers (in Hebrew). M.S. thesis, Tel-Aviv University.

Rasa, A. 1986. *Mongoose watch.* New York: Doubleday.

Redondo, T., and F. Castro. 1992. Signalling of nutritional need by magpie nestlings. *Ethology* 92: 193–204.

Reyer, H.U. 1990. Pied kingfishers: ecological causes and reproductive consequences of cooperative breeding. In *Cooperative Breeding in Birds,* ed. P.B. Stacey and W.D. Koenig, pp. 527–559. Cambridge, UK: Cambridge University Press.

Rhisiart, A.P. 1989. Communication and antipredator behaviour. D.Phil. thesis, University of Oxford.

Ridley, M. 1981. How the peacock got his tail. *New Sci.* 91: 398–401.

Ritland, D.B. 1991a. Unpalatability of viceroy butterflies (*Limenitis archippus*) and their purported mimicry models, Florida queens (*Danaus gilippus*). *Oecologia* 88: 102–8.

———. 1991b. Revising a classic butterfly mimicry scenario: Demonstration of Muellerian mimicry between Florida viceroys (*Limenitis archippus floridensis*) and queens (*Danaus gilippus berenice*). *Evolution* 45: 918–34.

Rohwer, S., and P.W. Ewald. 1981. The cost of dominance and advantage of subordination in a badge signaling system. *Evolution* 35: 441–54.

Rohwer, S., and F.C. Rohwer. 1987. Status signalling in Harris' sparrows: Experimental deceptions achieved. *Anim. Behav.* 26: 1012–22.

Roper, T. 1986. Badges of status in avian societies. *New Sci.* 109: 38–40.

Roseler, P.F., and C.G.J. Honk. 1989. Caste and reproduction in bumblebees. In *Social insects: An evolutionary approach to caste and reproduction,* ed. W. Engels, 147–166. Berlin: Springer-Verlag.

Rosenberg, E. 1984. *Myxobacteria.* New York: Springer-Verlag.

Rothstein, S.I. 1990. A model system for coevolution: avian brood parasitism. *Ann. Rev. Ecol. Syst.* 21: 481–508.

Rowel, T.E. 1962. Agonistic noises of the rhesus monkey (*Macaca mulata*). *Proc. R. Soc. London* 138: 91–96.

Rowley, I., and E. Russel. 1989. Splendid fairy-wren. In *Lifetime reproduction in birds,* ed. I. Newton, 233–52. London: Academic Press.

———. 1990. Splendid fairy-wren: Demonstrating the importance of longevity. In *Cooperative breeding in birds,* ed. P.B. Stacey and W.D. Koenig, 1–30. Cambridge, UK: Cambridge University Press.

Ryan, M.J. 1990. Sexual selection sensory systems, and sensory exploitation. *Oxford Surv. Evol. Biol.* 7: 157–95.

Ryan, M.J., M.D. Tuttle, and A.S. Rand. 1982. Bat predation and sexual advertisement in a neo-tropical anuran. *Amer. Nat.* 119: 136–39.

Sade, D.S. 1972. A longitudinal study of social behavior of Rhesus monkeys. In *The functional and evolutionary biology of primates,* ed. R. Tuttle, 378–98. Chicago: Aldine-Atherton.

Saino, N., A.P. Moller, and A.M. Bolzern. 1995. Testosterone effects on the immune system and parasite infestations in the barn swallow (*Hirundo rustica*): An experimental test of the immunocompetence hypothesis. *Behav. Ecol.* 6: 397–404.

Salyers, A.A., and D.D. Whit. 1994. *Bacterial pathogenesis.* Washington, DC: American Society for Microbiology.

Scherer, R.K. 1979. Nonlinguistic vocal indicators of emotion and psychopathology. In *Emotions in personality and psychopathology,* ed. C.E. Izard, 493–525. New York: Plenum.

————. 1985. Vocal affect signalling: A comparative approach. *Advances in the study of Behav.* 15: 189–244.

Schjelderup-Ebbe, T. 1922. Beiträge zur Sozialpsychologie des Haushuhns. *Z. Psychol.* 88: 225–52.

Schneider, D. 1992. 100 years of pheromone research. *Naturwissenschaften* 79: 241–50.

Schwammberger, K. 1993. Freilandbeobachtungen zur Nestübernahme bei *Polistes biglumis bimaculatus* durch den Sozialparasiten *Sulcopolistes atrimandibularis* (Hymenoptera, Vespidae). *Zeitschrift für Angewandte Zoologie* 79: 291–97.

Scott, D.H. 1988. Reproductive success in Bewick's swan. In *Reproductive success*, ed. T.H. Clutton-Brock, 220–36. Chicago: University of Chicago Press.

Selander, R.K. 1965. On mating systems and sexual selection. *Amer. Nat.* 99: 129–41.

————. 1972. Sexual selection and dimorphism in birds. In *Sexual selection and the descent of man 1871–1971*, ed. B. Campbell, 180–230. Los Angeles: University of California Press.

Sever, Z., and H. Mendelssohn. 1988. Copulation as a possible mechanism to maintain monogamy in porcupines, *Hysterix indica. Anim. Behav.* 36: 1541–42.

Seyfarth, R.M., D.L. Cheney, and P. Marler. 1980. Vervet monkey alarm calls: Semantic communication in a free-ranging primate. *Anim. Behav.* 28: 1070–94.

Shaulsky, G., and W.F. Loomis. 1993. Cell type regulation in response to expression of ricin A in *Dictyostelium. Dev. Biol.* 160: 85–98.

————. 1995. Mitochondrial DNA replication but no nuclear DNA replication during development of *Dictyostelium. Proc. Natnl. Acad. Sci.* 92: 5660–63.

Shepher, J. 1983. *Incest: A biosocial view.* New York: Academic Press.

Sherman, P.W., J.U.M. Jarvis, and R.D. Alexander. 1991. *The biology of the naked mole-rat.* Princeton: Princeton University Press.

Sigmund, K. 1993. *Games of life.* Oxford: Oxford University Press.

Simpson, M.J.A. 1968. The display of the Siamese fighting fish *(Betta splendens). Anim. Behav. Mon.* 1: 1–73.

Slagsvold, T. 1984. The mobbing behaviour of the hooded crow, *Corvus corone cornix:* Anti-predator defence or self-advertisement? *Fauna Norv.* Ser. C 7: 127–31.

————. 1985. The mobbing behaviour of the hooded crow, *Corvus corone cornix,* in relation to age, sex, size, season, temperature and the kind of enemy. *Fauna Norv.* Ser. C 8: 9–17.

Smith, H.G., and R. Montgomerie. 1991. Sexual selection and the tail ornaments of North American barn swallows. *Behav. Ecol. Sociobiol.* 28: 195–201.

Smith, H.G., R. Montgomerie, T. Poldmaa, B.N. White, and P.T. Boag. 1991. DNA fingerprinting reveals relation between tail ornaments and cuckoldry in barn swallows, *Hirundo rustica. Behav. Ecol.* 2: 90–98.

Smith, J.N.M., P. Arcese, and I.G. McLean. 1984. Age, experience, and enemy recognition by wild song sparrows. *Behav. Ecol. Sociobiol.* 14: 101–6.

Smith, N.G. 1966. Evolution of some arctic gulls *(Larus):* An experimental study of isolating mechanisms. *Ornith. Monogr. (AOU)* 4: 1–99.

————. 1967. Visual isolation in gulls. *Sci. Am.* 217: 94–102.

Smythe, N. 1970. On the existence of "pursuit invitation" signals in mammals. *Amer. Nat.* 104: 491–94.

Snow, D.W. 1976. *The web of adaptation: Bird studies in the American tropics.* New York: New York Times Books.

Snyder, H.S., and D.S. Bredt. 1992. Biological roles of nitric oxide. *Sci. Am.* 266: 58–67.

Soler, M., J. Soler, J.G. Martinez, and A.P. Moller. 1995. Magpie host manipulation by great spotted cuckoos: Evidence for an avian mafia? *Evolution* 49: 770–75.

Sordahl, T.A. 1990. The risks of avian mobbing and distraction behavior: An anecdotal review. *Wilson Bull.* 102: 349–352.

Spiro, M.E. 1963. *Kibbutz, adventure in utopia.* New York: Schocken Books.

Stacey, P.B., and W.D. Koenig. 1990. *Cooperative breeding in birds: Long-term studies of ecology and behaviour.* Cambridge, UK: Cambridge University Press.

Strahl, S.D. 1988. The social organization and behaviour of the hoatzin, *Opisthocomus hoatzin,* in central Venezuela. *Ibis* 130: 483–501.

Streeter, L.A., R.M. Krauss, V. Geller, C. Olson, and W. Apple. 1977. Pitch change during attempted deception. *J. of Personality and Soc. Psychol.* 35: 345–50.

Sugiyama, Y. 1967. Social organisation of hanuman langurs. In *Social communication among primates,* ed. S.A. Altman. Chicago: University of Chicago Press.

Sutter, E. 1994. Are the peacock's train feathers really upper tail coverts? *J. für Ornithol.* 135: 57.

Taborsky, M. 1994. Parasitism and alternative reproductive strategies in fish. *Adv. in Behav. Studies* 28: 1–100.

Tengo, J. 1984. Territorial behaviour of the kleptoparasite reduces parasitic pressure in communally nesting bees. *Abstract volume: XVII International Congress of Entomology* 510.

Thornhill, R. 1992a. Female preference for the pheromone of males with low fluctuating asymmetry in the Japanese scorpionfly (*Panorpa japonica,* Mecoptera). *Behav. Ecol.* 3: 277–83.

———. 1992b. Fluctuating asymmetry, interspecific aggression and male mating tactics in two species of Japanese scorpionflies. *Behav. Ecol. Sociobiol.* 30: 357–63.

Thornhill, R., and J. Alcock. 1983. *The evolution of insect mating systems.* Cambridge, MA: Harvard University Press.

Tinbergen, N. 1953. *The herring gull's world.* London: Collins.

———. 1965. Some recent studies of the evolution of sexual behavior. In *Sex and behavior,* ed. F.A. Beach, 1–33. New York: Wiley.

Todt, D., and H. Hultsch. 1995. Acquisition and performance of song repertoires: Ways of coping with diversity and versatility. In *Ecology and evolution of acoustic communication in birds,* ed. D. Kroodsma and E. Miller. Ithaca, NY: Cornell University Press.

Trivers, R. 1971. The evolution of reciprocal altruism. *Quart. Rev. Biol.* 46: 35–57.

———. 1974. Parent-offspring conflict. *Amer. Zool.* 14: 249–64.

———. 1985. *Social evolution.* Menlo Park, CA: Benjamin/Cumming Publication Company.

Trivers, R.L. 1972. Parental investment and sexual selection. In *Sexual selection and the descent of man 1871–1971,* ed. B. Campbell, 136–179. London: Heinemann.

Ulloa-Aguirre, A., A.R. Midgley Jr., I.Z. Beitins, and V. Padmanabhan. 1995. Follicle-stimulating isohormones: characterization and physiological relevance. *Endochrine Reviews* 16: 765–787.

Van der Blom, J. 1986. Reproductive dominance within colonies of *Bombus terrestris. Behaviour* 97: 37-49.

Van Lawick, H., and J. Van Lawick-Goodall. 1970. *Innocent killers.* London: Collins.

Van Lawick-Goodall, J. 1971. *In the shadow of man.* London: Collins.

Van Rhijn, J.G. 1973. Behavioural dimorphism in male ruffs (*Philomachus pugnax* L.). *Behaviour* 47: 153–229.

Veblen, T. 1899. *The theory of the leisure class.* New ed., ed. C.W. Mills. New Brunswick, NJ: Transaction Publishers, 1992.

Velthuis, H.W. 1990. Chemical signals and dominance communication in the honeybee *Apis mellifera* (Hymenoptera: Apidae). *Entomol. Gener.* 15(2): 83–90.

Wagner, R. 1992. Extra-pair copulations in a lek: The secondary mating system of monogamous razorbills. *Behav. Ecol. Sociobiol.* 31: 63–71.

Wagner, R., M.D. Schug, and E. Morton. 1996. Confidence in paternity, actual paternity and paternal effort by purple martins. *Anim. Behav.,* in press.

Wagner, R. H., 1996. Why do female birds reject copulations from their mates? *Ethology* 102: 465–480.

Wallace, A.R. 1889. *Darwinism: An exposition of the theory of natural selection with some of its applications.* London: Macmillan.

Ward, P. 1965. Feeding ecology of the black-faced dioch (*Quelea quelea*) in Nigeria. *Ibis* 107: 173–214.

———. 1972. The functional significance of mass drinking flights by sandgrouse (Pteroclididae). *Ibis* 114: 533–36.

Ward, P., and A. Zahavi. 1973. The importance of certain assemblages of birds as "information-centers" for food-finding. *Ibis* 115: 517–34.

Watson, J.P., and R. Thornhill. 1994. Fluctuating asymmetry and sexual selection. *Tree* 9: 21–25.

Weiner, J. 1994. *The beak of the finch.* New York: Vintage.

West, M.J. 1969. The social biology of polistine wasps. *Univ. Mich. Mus. Zool. Misc. Publ.* 140: 1–101.

West-Eberhard, M.J. 1975. The evolution of social behavior by kin selection. *Quar. Rev. Biol.* 50: 1–33.

———. 1984. Sexual selection, competitive communication and species-specific signals in insects. In *Insect communication,* ed. T. Lewis, 283–324. New York: Academic Press.

———. 1986. Dominance relations in *Polistes canadensis* (L.), a tropical social wasp. *Monitore Zool. Ital.* (N.S.) 20: 263–81.

Wiklund, C., and T. Jarvi. 1982. Survival of distasteful insects after being attacked by naive birds: A reappraisal of the theory of aposematic coloration evolving through individual selection. *Evolution* 36: 998–1002.

Wilhelm, K. V., H. Comtesse, and W. Pflumm. 1980. Zur Abhängigkeit des Gesangs vom Nahrungsangebot beim Gelbbauchnektarvogel (*Nectarinia venusta.*) *Z. Tierpsychol.* 54: 185–202.

———. 1982. Influence of sucrose solution concentration on the song and courtship behaviour of the male yellow-bellied sunbird (*Nectarinia venusta*). *Z. Tierpsychol.* 60: 27–40.

Williams, G.C. 1966. *Adaptation and natural selection: A critique of some current evolutionary thought.* Princeton: Princeton University Press.

———. 1994. *Natural selection: Domains, levels and challenges.* Oxford: Oxford University Press.

Willson, D.S., and E. Sober. 1994. Reintroducing group selection to human behavioral sciences. *Behavioral and Brain Sciences,* 17: 585–654.

Wilson, E.O. 1971. *The insect societies.* Cambridge: Harvard University Press.

————. 1975. *Sociobiology: The new synthesis.* Cambridge, MA: Belknap Press.

Winston, M.L., and K.N. Slessor. 1992. The essence of royalty: Honey bee queen pheromone. *Amer. Sci.* 80: 374–85.

Witherby, H.F., F.C.R. Jourdain, N.F. Ticehurst, and B.W. Tucker. 1943. *The handbook of
British birds.* London: Witherby.

Woolfenden, G.E., and J.W. Fitzpatrick. 1984. *The Florida scrub jay.* Princeton: Princeton
University Press.

————. 1990. Florida scrub jays: A synopsis after 18 years of study. In *Cooperative Breeding
in Birds,* ed. P.B. Stacey and W.D. Koenig, 239–66. Cambridge: Cambridge University
Press.

Wyllie, I. 1975. Study of cuckoos and reed warblers. *Br. Birds* 68: 369–78.

Wynne-Edwards, V.C. 1962. *Animal dispersion in relation to social behaviour.* Edinburgh:
Oliver and Boyd.

————. 1986. *Evolution through group selection.* Oxford: Blackwell Scientific Publications.

Yom-Tov, Y. 1989. Seemingly maladaptive behaviours, individual recognition and hierarchy. *Ornis Scand.* 20: 2.

Zahavi, A. 1971a. The function of pre-roost gatherings and communal roosts. *Ibis* 113:
106–09.

————. 1971b. The social behavior of the white wagtail, *Motacilla alba,* wintering in Israel.
Ibis 113: 203–11.

————. 1974. Communal nesting by the Arabian babbler, a case of individual selection.
Ibis 116: 84–87.

————. 1975. Mate selection: A selection for a handicap. *J. Theor. Biol.* 53: 205–14.

————. 1976. Cooperative nesting in Eurasian birds. In *Acta XVI Con. Inter. Ornithol.,* ed.
H.J. Frith, and J.H. Calaby, 685–93. Canberra, Australia.

————. 1977a. Reliability in communication systems and the evolution of altruism. In *Evolutionary ecology,* ed. B. Stonehouse and C.M. Perrins, 253–59. London: Macmillan
Press.

————. 1977b. The cost of honesty (further remarks on the handicap principle). *J. Theor.
Biol.* 67: 603–5.

————. 1977c. The testing of a bond. *Anim. Behav.* 25: 246–47.

————. 1978a. Decorative patterns and the evolution of art. *New Sci.* 80: 182–84.

————. 1978b. Why shouting. *Amer. Nat.* 113: 155–56.

————. 1979. Parasitism and nest predation in parasitic cuckoos. *Amer. Nat.* 113: 157–59.

————. 1980. Ritualization and the evolution of movement signals. *Behaviour* 72: 77–81.

————. 1981a. Natural selection, sexual selection and the selection of signals. In *Evolution
today,* ed. G.G.E. Scudder and J.L. Reveal, 133–38. Pittsburgh: Carnegie-Mellon University Press.

————. 1981b. The lateral display of fishes: Bluff or honesty in signalling? *Behav. Anal.
Lett.* 1: 233–35.

————. 1981c. Some comments on sociobiology. *Auk* 98: 412–14.

————. 1982. The pattern of vocal signals and the information they convey. *Behaviour* 80:
1–8.

————. 1983. This week's citation classic: The importance of certain assemblages of birds
as "information centers" for food finding. *Current Contents* 15: 26.

————. 1985a. Evolution of group life with special reference to the Arabian babbler (*Turdoides squamiceps*). In *Acta XVIII Con. Inter. Ornithol.,* ed. V.D. Ilychev and V.M. Gavrilov 2: 1043. Moscow: Academy of Sciences of the USSR.

————. 1985b. Some further comments on the gatherings of birds. In *Acta XVIII Con. Inter. Ornithol.,* ed. V.D. Ilychev and V.M. Gavrilov, 2: 919–20. Moscow: Academy of Sciences of the USSR.

————. 1987. The theory of signal selection and some of its implications. In *Proc. Intern. Symp. Biol. Evol.,* ed. V.P. Delfino, pp. 305–27. Bari, Italy: Adriatica Editrica.

————. 1988a. Mate guarding in Arabian babbler, a group-living songbird. *Acta XIX Con. Inter. Ornithol.,* 420–227. Ottawa.

————. 1988b. Mate choice through signals. *Acta XIX Con. Inter. Ornithol.,* 956–60. Ottawa.

————. 1989. Arabian babbler. In *Lifetime reproduction in birds,* ed. I. Newton, 253–76. London: Academic Press.

————. 1990. Arabian babblers: The quest for social status in a cooperative breeder. In *Cooperative breeding in birds: Long-term studies of ecology and behaviour,* ed. P.B. Stacey and W.D. Koenig, 103–30. Cambridge: Cambridge University Press.

————. 1991a. On the definition of sexual selection, Fisher's model, and the evolution of waste and of signals in general. *Anim. Behav.* 42: 501–3.

————. 1991b. Sexual selection: Badges and signals. *Tree* 7: 30–31.

————. 1993. The fallacy of conventional signalling. *Phil. Trans. R. Soc. Lond. B* 338: 227–30.

————. 1995. Altruism as a handicap—the limitations of kin selection and reciprocity. *Avian Biol.* 26: 1–3.

————. 1996a. The evolution of communal roosts as information centers and the pitfall of group-selection: A rejoinder to Richner and Heeb. *Behav. Ecol.* 7: 118–19.

————. 1996b. Cooperation among lions: An overlooked theory. *Tree* 11: 252.

Zahavi, A., and D. Ralt. 1984. Social adaptations in myxobacteria. In *Myxobacteria,* ed. E. Rosenberg, 215–20. New York: Springer-Verlag.

Zahavi, T. 1975. Social behaviour of the babblers (in Hebrew). High school project.

Zilberman, R. 1991. Extra-pair copulations in a sunbird population (*Nectarinia osea osea*) in Ramat-Aviv area (in Hebrew). M.S. thesis, Tel-Aviv University.

Zuk, M., K. Johnson, R. Thornhill, and D. Ligon. 1990. Mechanisms of female choice in red jungle fowl. *Evolution* 44: 477–85.

LIST OF FIGURES

INDEX